Praise for
How Stella Learnea

"A talented speech therapist who works ... tina Hunger trained her puppy to use words in a meaningful way. She was amazed that language learning in her dog was similar to that of a young child. The results speak for themselves. A wonderful book."

—Temple Grandin, author of *Animals in Translation*

"Christina Hunger presents an entertaining and practical guide to embarking on the incredible journey of interspecies communication with her dog. When we are open to the idea that communicating with another species is possible, and provide a language-enriched environment, we create the space for shared language and a deep connection. Christina has provided the key that unlocks this potential for all of us."

—Penny Patterson, PhD, president and research director of The Gorilla Foundation and Project Koko

"Humans and dogs have been evolving together and forming an extremely unique and special bond for thousands of years. As anyone who's known and loved a dog will agree, the natural ability that dogs have to understand and communicate with us is truly incredible, going much deeper than just wagging tails and happy barks when they hear the word 'walk.' I'm so thrilled that Christina is using her experience as a speech and language therapist to give dogs a way to 'speak' with us in a more complex, profound way."

—Zak George, host of Zak George's Dog Training Revolution

"[A] fascinating study of the untapped potential in human-dog interaction."

—*Booklist*

"This delightful book is a joy to read. But it also has an important message for the study of animal languages: When you think outside the box, you can come up with amazing results. Christina's use of augmentative and alternative communication (AAC) with her dog, Stella, employing techniques that are taught to children, has shown that a dog has the cognitive capability to learn to use buttons as words and to formulate meaningful sentences. She also has very practical advice on how to teach your dog to do the same. I highly recommend this book."

—Con Slobodchikoff, PhD, author of *Chasing Doctor Dolittle: Learning the Language of Animals*

"A must-read for anyone who ever wished their dog could talk, Stella's story will lift your spirits and inspire your soul. Drawing on her expertise in language development and augmentative communication, speech-language pathologist Christina Hunger takes us on an incredible journey of joy, discovery, and innovation that will forever change the relationship between dogs and their humans."

—Shari Robertson, PhD, CCC-SLP, 2019 president of the American Speech-Language-Hearing Association

"Over a year into the pandemic, your bookshelves may be so crowded with books that you can't imagine adding another. But a three-year-old dog named Stella will have you rethinking that."

—*Chicago Tribune*

"The mixture of memoir and how-to guide strikes an effective balance for this dog mom. . . . If you've been thinking about trying to use recordable buttons with your dog, then this book will offer plenty of insights for you."

—McSquare Doodles

How Stella Learned to Talk

How Stella Learned to Talk

The Groundbreaking Story of the World's First Talking Dog

Christina Hunger

wm
WILLIAM MORROW
An Imprint of HarperCollins*Publishers*

The names and identifying details of some individuals depicted in this book have been changed to protect their privacy. This book is for informational and entertainment purposes only. No guarantees are made with respect to the information and advice contained herein. The publisher and the author assume no responsibility for any injuries, damages, or losses incurred as a result of following this advice.

HarperCollins books may be purchased for educational, business, or sales promotional use. For information, please email the Special Markets Department at SPsales@harpercollins.com.

A hardcover edition of this book was published in 2021 by William Morrow, an imprint of HarperCollins Publishers.

FIRST WILLIAM MORROW PAPERBACK EDITION PUBLISHED 2022.

Library of Congress Cataloging-in-Publication Data has been applied for.

ISBN 978-0-06-304684-9

22 23 24 25 26 LSC 10 9 8 7 6 5 4 3 2 1

For my husband, Jake.
I couldn't have done any of this without you.
I love you so much.

Contents

How Stella Learned to Talk

Prologue

"Bye, Stella," I said while I ate breakfast at the dining table. "Have fun with Jake."

My fiancé was holding Stella's leash as he waited by the front door. They left together every morning to play at the beach or park while I got ready for work.

"Ready, Stella?" Jake asked.

Stella paused in the kitchen. She turned to look at the door, then she locked eyes with me. I wondered what she was thinking. Normally, she hurries out the door in the morning.

Stella approached her communication board on the floor. It was two feet by four feet, filled with colorful buttons, each programmed with a prerecorded word. She pressed four different buttons in a row.

"Christina come play love you," Stella said. She hopped off her board and stared at me again.

Jake laughed, and I couldn't help but smile. "You want me to come play, Stella?"

She wagged her tail.

I threw on shoes and a coat, then grabbed the leash from Jake.

My dog just invited me to come play with her. How special is that?

CHAPTER 1

Presume Competence

As I sat on a six-foot-wide swing resembling a giant padded block, I held my breath in anticipation. Oliver, my speech therapy client, was sitting with me holding his tablet-sized communication device. Oliver rarely initiated using his talker without my help. But today he grabbed it as soon as we met. *What did he want to tell me?*

Even though Oliver was only nine years old, he was almost as tall as me, his twenty-four-year-old speech-language pathologist. His basketball shorts and T-shirt made him look like any typical nine-year-old boy. But the braces on his legs, noise-reducing headphones covering his ears, and communication device strapped over his shoulder indicated that something was different. Oliver had autism spectrum disorder. He had been coming to this pediatric clinic in Omaha, Nebraska, for physical therapy, occupational therapy, and speech therapy for years.

Oliver squinted, his finger hovering over the screen. I focused on what Oliver was about to say, blocking out the noise of children swinging, climbing the rock wall, and riding on scooters around us in the gym.

Possibilities ran through my head. Maybe he would say one of the words we had been practicing on the swing for the past couple of weeks. Since Oliver learned best while he was in motion, we spent much of our sessions practicing words such as *go*, *stop*, *fast*, and *slow* while swinging. I loved modeling the word *fast* on Oliver's device and pushing him as high as I could. Oliver would squeal and throw his head back in laughter, savoring every second of the increased speed. It was impossible for anyone to watch him without smiling. His pure joy lit up the entire room.

Oliver tapped one of the sixty icons on his tablet, which opened a new page of word choices. He traced the brightly colored rows with his index finger before pausing over a single square. When Oliver pressed it, the synthesized voice said, "rice."

"Rice?" I paused. Oliver and I were in the middle of a gym, not a kitchen. "Rice" was about the last thing I expected him to say. "Eat rice at lunch?" I asked. After I said each word, I pushed the corresponding button on his tablet. The best way for Oliver to learn how to use his talker was to see other people using it as well.

Oliver grunted and kicked his legs in frustration. That was not what he was trying to say.

I glanced at my watch, 4:35. "Maybe you want rice for dinner after therapy," I said. "Want eat rice?"

Oliver swatted my hand away, then repeated himself, "Rice rice."

Oliver said "rice" in our session last week, too, but I did not really think anything of it. Since there was no rice around, I chalked it up to Oliver exploring new vocabulary. He had only had this communication device for a few months, and it can take a while before kids start using words intentionally on their own. They need time to explore words and to see and hear their communication system in use before they can be expected to talk with it. It's like how babies need to hear people using language for a whole year before they start saying words on their own. Now that I knew Oliver saying "rice" was not an isolated occurrence, I modeled everything I could think of that was related to rice. I used Oliver's device to try a variety of phrases.

Was Oliver trying to tell me he was hungry and wanted to eat rice? Was he trying to say he liked rice? What if he hates rice and was trying to tell me it is bad? I modeled "rice good," "rice bad," "like eat rice," "all done eat rice," "Oliver eat rice," "school eat rice," everything I could possibly think of. Each time I tried a new phrase, I paused to assess his reaction. Oliver understood much more than he could say; his receptive language skills were significantly greater than his expressive language skills. If I said the right phrase he was trying to communicate, he would get excited. It reminded me of how I felt when trying to say a word in French years after I learned it in high school. If I read the word or heard someone say it, I would know what it meant. But it was so much harder for me to come up with the word on my own. That's because my French receptive language skills were much higher than my French expressive language skills.

Oliver ripped his headphones off and threw them to the

ground. He rattled the swing, knocking my clipboard off it while grunting.

"It's okay, it's okay. We're okay. You sound mad, Oliver," I said. I held the device in front of him and modeled, "mad." I needed to redirect him before things got worse. The fastest way to do that was to start swinging. No matter what was going on, Oliver always loved to swing. This is true for a lot of children with autism. The rocking motion helped regulate his sensory system. Within a minute of using all my body weight to propel Oliver as high as possible, he returned to his giggly self. I fixed the situation for now so we could salvage our session but did not truly solve the problem. *Why is Oliver saying "rice"? And why is Oliver getting so mad when I try talking about rice?*

Oliver was one of the many children on my caseload learning how to use a communication device to talk. Communication devices are a form of augmentative and alternative communication (AAC). AAC is a fantastic tool that gives people with severe speech delays or disorders the ability to say words through another medium. Occasionally Oliver verbally repeated single words he heard in a video or from whoever around him was speaking, but at nine years old, that was the extent of his verbal speech abilities. Many people mistakenly believe that verbal speech skills represent a person's intelligence or language abilities. One of my favorite things about being a speech-language pathologist is shattering this misconception. I love introducing AAC to children who had been misunderstood for years and watching great transformations unfold.

In graduate school, one of my clients had been using an AAC device for a couple of years. She primarily communicated in two-word phrases until the day she felt the sleeve of a fuzzy sweater I had not worn there before. She looked at me and said,

"New purple sweater like." I had no idea that she knew how to say the word *sweater* or had been paying attention to my wardrobe this whole time.

At the clinic, I recently increased another one of my clients' number of visible words on his device from about twenty to the full capacity of two thousand words. On the same day that I made the change, the child said "down stairs down stairs green block" then bolted to the door. He pulled as hard as he could on the doorknob but did not have the dexterity to turn it. If he had not said what he wanted, I would not have opened the door. I would have assumed he was trying to escape from the session. In the therapy area downstairs, he walked straight to the game cabinet and pulled out a cardboard box from the middle of the shelf, not even fazed by all the games he knocked down in the process. He flipped open the lid and picked out all the green magnetic shapes, then started stacking them in different ways. Before today, *down* would have been the only word of that phrase he could have said. How long had he been wanting to tell me the exact game he wanted to play? How long had he been pulling on doorknobs trying to lead me to what he wanted, not to try running away? I could not wait to share this great breakthrough with his mom at the end of the session.

Some parents, and even some professionals, assume that kids with significant disabilities know very little or are too difficult to teach. In my experience, many professionals just didn't know how to give them the opportunity to learn and discover their potential or didn't stick something out long enough to give it a real chance. For all children, belief in their potential makes a huge difference in their learning. But for kids using AAC, belief in potential is everything. If the therapist or parent does not truly believe in the child's potential, they physically limit what is possible for the child to say by only giving them a few options. When professionals set the bar low with AAC, the child will be stuck.

This is what happened with Oliver before I took over his case. His communication device was programmed to only say about eight or ten different sentences. He could push one icon that said, "I want more swing" or push another that said, "I need to go to the bathroom." The therapist before me thought that was all he could handle. For entire thirty-minute-long therapy sessions, Oliver would say "I want more swing" repeatedly. I could not blame him, though; he did not really have any other options. He could not say anything else about the swing or talk about any other activities he might participate in throughout his typical day. Oliver did the best he could with what he had.

After getting to know him more and gaining my own confidence, I set up a conference with his mom to show her how Oliver's device could be programmed to have over two thousand words for him to learn to say instead of the same old ten sentences. This was actually the language program's default setting, how it was supposed to be used. His prior therapist went out of her way to remove all the words and replace them with sentences she created. With access to so many words, Oliver could eventually create millions of different sentences, like we all have the privilege to do. But this was only months into my first year as a speech-language pathologist. I had never made such drastic changes to a child's communication system or scrapped a more experienced speech therapist's treatment plan to create my own. In graduate school, we learned how to set up AAC devices and make solid treatment plans from the start. Fixing ineffective plans was a new territory.

I prepared for the meeting, looking over my notes, pulling together articles on the importance of single-word-based AAC systems instead of preprogrammed sentences, communicating for functions other than requesting, and teaching words that could be used in a variety of contexts rather than focusing only on saying nouns. I hoped his mom could not sense how nervous I was. It was important for her to believe me. Oliver could be learning so much more than anyone had given him the chance to say.

"He's done such a great job with what he has," I said to her as we sat together in a small conference room. "He needs more words, though. I know it will be a transition, but Oliver can learn to say so much more than what he has available to him right now." I showed her the AAC articles as I spoke. "This system will set him up for a lifetime of communication opportunities and allow him to grow," I said. "Think about when your daughter was learning to talk. She babbled first, then said words, then started putting words together, and eventually formed her own sentences, right?"

When she looked up at me, I saw a hesitant mother who had been let down by the years of therapy that did not work for her son. This could not have been easy for her. Oliver already had several speech therapists before me, who each had their own visions for his communication progress. It must have been confusing and exhausting to hear so many different professional opinions and have to figure out which made the most sense for her son. "Yes," she said.

"Oliver needs to go through those stages of language development, too, just in a different way."

"I get it. This looks awesome," she said, tapping the pages. "I'm just . . . I'm just nervous. Oliver has been using his device for a while now and his school knows how to use it. We're all used to it."

I nodded. "We can definitely save this language set in case we ever want to go back. And I'll reach out to his speech therapist at school so we can all be on the same page about targeting the same types of words and concepts. She can let me know what's important at school, and I can let her know what we're working on here. I think it will be worth it."

She asked some other questions, then gave me the green light to move ahead. I was thrilled with the idea that Oliver's world would open up to him, not to mention honored his mom placed her trust in me. In that moment, I committed myself to not only helping Oliver be the best communicator he could be, but also to not letting her down. I adjusted the settings on the device and marveled at all the new words we could use in our sessions.

Oliver progressed through stages of learning as I expected he would. First, he spent time "babbling" on his device by pushing random buttons and observing what would happen. He played with his words just like babies play with sounds as they are learning to talk. Then, Oliver quickly started reusing words he already knew such as *more*, *swing*, and *go* in appropriate situations. Now, Oliver was starting to say new words like *turn* when he wanted me to spin the swing round and round, or *stop* when he wanted to be finished with an activity. This was the first time in his life that Oliver could tell someone to stop by saying a word instead of by screaming, kicking, or scratching. Oliver was not an aggressive child. He was only doing what he could to convey his message. Oliver's other therapists, his mom, and I were all starting to see that Oliver had all these opinions that he had no way to express. I thought Oliver was really getting the hang of the new device setup.

Until today and "rice," everything he said was making sense. I dropped Oliver off at his physical therapist's desk, where I shared which words we targeted today. His physical therapist, occupational therapist, and I all tried to incorporate one another's goals whenever possible. I popped into the waiting room to talk with his mom. Debriefing at the end of each appointment was crucial in putting the pieces of Oliver's communication together. After one session, for example, Oliver's mom's eyes grew when I casually mentioned that he said "dentist" when we were playing a game. She explained that Oliver left school early that day to go to the dentist. Learning that was huge for me. Without this context, I would have completely missed that he was trying to tell me about his day. Using words to share information is an exciting language milestone. This communication function would have been literally impossible for him to achieve with the limited sentences previously programmed into his device. Oliver's mom was so grateful, and we were both thrilled. Oliver could say much more than what he wanted or needed in the moment.

I crouched down next to her in the crowded waiting room and said, "Has Oliver been saying 'rice' at home at all?"

She frowned. "No, not that I can remember. Why?"

"He said it to me this week and last week. Today he was getting pretty upset when I tried figuring out what he meant. Does he eat rice for lunch or dinner?" I said.

"I mean we have it sometimes, but I don't think he has any strong feelings toward it. He's just . . . he's just Oliver." She sighed. "It could mean anything."

"I'm sure we'll figure it out. Let me know next week if you hear it at home at all," I said. "Also, before I forget, Oliver discovered the dinosaur page on his device. He was repeating dinosaur names over and over while we were cleaning up. We ran out of time, but next week I'll bring some dinosaur toys."

Over the course of the next three weeks, Oliver began each session the same way. He continued saying "rice," and I continued saying anything I could think of that had to do with rice. That word haunted me. I printed off pictures of rice, brought in a pretend box of rice from our play kitchen, studied all the other food words on his device to see if the icon looked similar to another. Maybe he was trying to say a different word, but could not find it? I checked in with his occupational therapist and physical therapist, too. Apparently, Oliver was only saying "rice" when he was with me, nobody else. *Why?* I wondered. Oliver also rejected all the different dinosaur toys and books I brought. He displayed absolutely no interest in any of them even though he continued repeating dinosaur names. I was completely flummoxed.

During these three weeks, Oliver became more and more upset when I responded to "rice." He started screaming louder, dug his fingernails into my forearm, desperate for me to understand him. He chucked his device on the ground, and even at me several times. I moved my head just in time for it to land on the mattress on the floor instead of hitting my face. He was doing everything in his power to let me know he was upset. Taking deep breaths and chanting to myself in my head, *He's acting how he's feeling, he's acting how he's feeling*, kept me from losing my patience. Graduate school prepared me for many aspects of my job, but this was not one of them. What was I doing wrong?

Before this "rice" saga, Oliver and I had a great relationship.

He did not always get along well with adults, and he clearly trusted me. We connected through little games that became inside jokes he anticipated every week. Oliver loved directing me to say specific animal sounds. It cracked him up to hear me shout, “Cock-a-DOODLE-DOOOO” after he said “rooster” or to hear me howl at the ceiling after he said “wolf.” If I didn’t make the animal noise, Oliver would gently tap my arm to remind me of my role in our game. Oliver and I got along so well because I did not try to force him to do what *I* wanted. Instead, I tried as best I could to follow his interests and expand on them. It was not my job to make Oliver say certain things at certain times. It was my job to give him the skills to talk about anything he wanted to talk about and connect with important people in his life.

Now seeing Oliver this upset during our sessions scared me. He was a strong kid. I had no idea what would happen next if this pattern continued. Would he hurt himself? Would he hurt me? It was heartbreaking to see this complete shift in his behavior. I almost didn’t recognize him anymore. Our communication and rhythm had broken down. I feared I was losing his trust more and more each week. The longer this continued, the more difficult it would be to gain it back.

The following Saturday morning, my first client of the day was a sweet elementary-aged girl named Anna who always arrived right on time in her pajamas. Each week, she would immediately tell me where she wanted to play that day. Ninety percent of the time, she asked to go to the gym. If I were her age, I would also prefer to be in a room filled with mattresses on the floor to bounce on, ten different swings, a pillow pit to jump in, and a rock wall to climb. The open play area and sensory gym were very popular with the kids.

That morning she asked, "Can we play upstairs?" That was fine with me. We picked out a book and a game from the shelves and went upstairs to the individual treatment rooms. We walked down the hall, flipping on the lights as we went. We must have been the first ones up there that morning.

"In here!" Anna shouted as she opened the second door on the right. She sighed with disappointment when she realized this room only had a child-size table and chairs, nothing exciting. It looked much less fun than the gym.

"This table will be perfect for our book and game! I'm glad we're up here this morning," I said. Anna looked skeptical. She walked around the table and found a small, clear bin on the floor.

"Miss Christina, what's that?" she asked.

"Good question! It's a bin filled with . . . oh my gosh." I jumped out of my chair.

"A bin what?"

"A bin filled with rice," I said.

She joined me on the ground, suddenly very curious about what we would find as I ripped the lid off. "Look, Miss Christina! Dinosaurs in here." She dove her hands through the rice to find the rest of the dinosaurs. "This feels nice!" Kids with sensory differences loved running their hands through the rice. It felt calm and soothing.

I shook my head and laughed. This must have been what Oliver was trying to tell me for the past five weeks. I had exhausted every other option. He did not want to eat rice. He did not want to look at dinosaur books or videos. No wonder he was getting so mad when I was not understanding. He wanted to play with the rice bin filled with toy dinosaurs.

When Oliver came back to the clinic for therapy on Tuesday, I felt like a little kid trying to keep a secret. I even went so far as to hide the bin behind a stack of cones in the gym. I so badly wanted to shout "*I figured it out! I know you've been asking for the rice bin with dinosaurs this whole time*" as soon as I saw him. Instead, I waited to see if Oliver would ask for it again. I led Oliver through the door to the sensory gym with a little more pep in my step than I typically had. Oliver sat down on the swing as usual, tapped his device to turn the screen on, and said, right on schedule, "rice."

"Rice! Yes! Let's play in the rice," I said.

Oliver's eyes followed me as I walked across the gym to pick up the rice bin I had hidden. By the time I reached the swing, Oliver was shrieking with joy, louder than I had ever heard him yell before. Oliver scooped the rice and threw it in the air and dug for the dinosaurs. I did not care how messy he was being; he deserved this celebration after he waited six weeks for me to understand him. In the thirty-minute session, Oliver only paused his ecstatic play once. He took his hand out of the rice bin and gently squeezed my forearm.

After I talked with Oliver's mom and other therapists about the conclusion of the "rice" saga, I learned that months ago Oliver used to play in the rice bin with the speech therapist who no longer worked at the clinic. That's why he was only asking me for it. Oliver's mom was shocked to learn that he was not saying random words for the past six weeks and getting mad for no reason. She looked at Oliver with a newfound pride and relief in her eyes. She and I knew, once and for all, that switching his device setup was the right decision. We had only caught a glimpse of Oliver's knowledge, perseverance, and potential and could not wait to hear what else he was trying to tell us for so long. In every session from there on out, I opened myself up to all the possibilities of what he could know and be trying to communicate. No words, phrases, or sentences were off-limits. I modeled everything I could for him. Oliver went on to create his own two- and three-word phrases every day, use adjectives like "mean" and "nice" to talk about the people in his life, "funny" after watching his favorite videos, and even told me "speech pathologist sick" after I returned to work from being out with the flu.

My graduate school professors instilled the importance of "presuming competence" when working with children who require AAC. Presuming competence means treating from the fundamental understanding that everyone can learn, and everyone has something to say. It is impossible to know a person's communication potential before they are given the tools and interventions they need to succeed. To everyone else, Oliver was a lost cause. He could only push buttons with complete sentences to express a few different desires. If he did not get what he wanted, he often screamed, kicked, and acted out. Underneath all these behaviors was a child who was so misunderstood, someone with so much potential. He just needed a chance to learn, the right tools, and someone to believe in him. It was impossible for me to fully grasp the significance of presuming competence sitting in a classroom. Now, seeing what happened when we gave Oliver a real chance to learn words and believed in him was life changing. I looked at him, and every child on my caseload, through a lens of potential. I never made assumptions about what any child might know, or be capable of learning. We need to give all kids, regardless of impairments, the opportunity to learn. We need to argue for their potential, not their limitations. I loved seeing what kids could say when given a real chance to learn. I loved the transformative power of AAC. I craved more of it.

CHAPTER 2

Ozzie and Truman

Ding ding ding. The bell hanging from the front doorknob chimed. Ozzie, one of the two goldendoodles that my boyfriend, Jake, and I were dog sitting, stared at us from the doormat.

Jake squatted down in front of Ozzie to scratch his white, shaggy fur. "You want to go outside again, bud? Weren't you just out?" Jake asked.

Ozzie lifted his paw and swatted at the bell again. He and his partner in crime, Truman, a black goldendoodle, had been trained to ring a bell when they needed to go outside. We hung the bell on our front doorknob to stay consistent with the setup at their house.

"Ozzie really loves ringing that bell. Maybe he wants to go for a walk," I said. "Want to take them together?" As soon as he heard the word *walk*, Truman hopped off the couch to join Jake and Ozzie near the door.

"That looks like a yes to me," Jake said. "Let's get your leashes on."

Omaha was beginning to break out of winter for the year. Mud piles from melted snow lined the edges of the front yards in our neighborhood. Ozzie and Truman panted with their tongues lolling out the sides of their mouths, happy to be out and about exploring a new neighborhood.

Ozzie and Truman belonged to my former speech therapy supervisor, Mandy. When I heard that Mandy and her family were going out of town for the weekend, I jumped at the opportunity to watch her two goofy dogs. I had dog sat them a few months before, and got a kick out of their different personalities. Ozzie went with the flow of what was happening and adjusted to changes pretty well. Besides ringing the bell when he wanted to go outside, he did not really demand much else from us. He just hung around keeping us company. Truman, however, was more energetic and anxious. He seemed a little unsure about everything, like he could not ever fully relax.

Jake was game to watch them, too. We had been dating for the past eleven months and had just moved in together. Originally from a small town in northern Minnesota, Jake moved to Omaha about five years ago to work as a financial analyst for a large agricultural company. When Jake was growing up, his family lived in the country and had a hundred-pound black lab named Kirby. Kirby spent the day outside exploring the woods behind Jake's house and slept in their insulated shed at night. He swam in the pond to chase fish and fetched baseballs Jake's family would hit deep into the forest for him to find. Taking dogs outside to go to the bathroom and to go on walks was new for Jake.

As a child I desperately wanted a dog, but my struggles with asthma prevented that. In lieu of a real-life furry companion, I collected countless stuffed dogs, robotic dogs, and Beanie Baby dogs throughout my childhood. When I was in second grade, my parents surprised me with a box turtle. I would walk Shellie in our front yard, watch her dig holes by my mom's hostas, feed her, give her baths, and pet her red speckled head whenever she emerged from her shell. No matter how much I treated her like a dog and no matter how many toy dogs I collected, though, my desire for a puppy never went away.

Thankfully, when I was ten years old, I received the official health clearance from my doctor to adopt a dog. I immediately started researching puppies and wrote a persuasive essay to convince my parents that we should have one. I had waited years for this possibility; I *had* to bring my parents on board. Shortly after my essay and presentation, we brought home the new member of our family, a sweet, playful boxer puppy named Wrigley.

Living with a dog was magical. This little, wiggly, slobbery creature joined our family and fit right in. Wrigley's intelligence left all of us in awe. Wrigley learned the patterns of when my older sisters would leave for college. As soon as she saw one of them walk a suitcase up from the basement, she stayed right by their side until their departure. Sometimes she would even lie in their suitcases while they packed. After they left, Wrigley always spent the rest of that day curled up in a ball on their beds. It seemed like her ritual to acknowledge that they would not be back for a while.

Wrigley also knew that my mom was the only consistent

enforcer of the "no dogs on couches" rule. If my dad was home with us, Wrigley would freely hop up to cuddle until she heard my mom's car pull up. Then Wrigley fled the scene and lay on her bed as if nothing had happened. We all laughed and cherished this secret between us. My sisters and I loved trying to convince Wrigley to join us on the couch when my mom was upstairs or in another room. We would encourage her by patting on the cushions. "Come on, girl, come on up." Wrigley looked around cautiously several times before determining it was all clear.

Wrigley always wanted to go play or sunbathe in the backyard. If we were watching TV, she would stand right in front of it whimpering until we let her out. I bet if she had a bell to ring when she wanted to go out like Ozzie and Truman did, we would have heard it all the time. She became so excited for her walk every night. As soon as my mom would grab her headphones and put her tennis shoes on, Wrigley would perk up and run to her leash hanging in the back hall. When my mom would finally ask if she wanted to go for a walk, Wrigley shook her whole body in excitement and whined until they left. She was so expressive and intelligent.

Wrigley had passed away a few months before we watched the goldendoodles. I had only been back to my parents' house in Aurora, Illinois, once or twice since she died and noticed a huge difference. For over thirteen years I was used to Wrigley greeting me with kisses every time I walked through the front door, snuggling next to me on my bed, and running through the house with a toy hanging out of her mouth. I did not realize how central Wrigley's role in our family was until I experienced the stillness of our house when she was gone.

When Jake and I and the goldendoodles all arrived back home, the four of us hopped on the couch, ready to relax.

Truman jumped off the couch and stood in the middle of the living room. "What's up, Truman?" I asked. Truman cried. He sounded like a scared child. "Hm. Let's see if you need water," I said. Truman followed me into the kitchen. "Nope, you have water. And we were just outside so you shouldn't need to go to the bathroom. And you ate all your breakfast so you shouldn't be hungry."

He had to be upset for a reason. What was he thinking about and trying to express? Did he want a certain toy that we did not have here with us? Did he want to be back home? Was he stressed?

Truman kept whimpering. *Are you missing your family?* I wondered. I squatted down to Truman's level, petting his face while staring in his eyes. "What's wrong, buddy?" Truman and I walked back to the living room. "I don't know what's going on. I hope he's okay," I said to Jake. I felt so bad for Truman. It is so frustrating to try communicating something and not be understood.

"Oh, he's fine. I bet he just wants to play!" Jake leapt from the couch and chased Truman up the stairs and back down. Before I knew it Jake, Ozzie, and Truman were running laps around the living room, kitchen, hallway, and dining room. Every time Jake turned a corner, the dogs slid across the hardwood floors. I stayed out of the way, enjoying the show, but it still nagged at me. While Jake had diverted them, I still didn't know what Truman really wanted.

The next morning, Jake and I let the dogs out in the backyard while we cooked pancakes.

"You know, this house is perfect for a dog," I said.

"What do you mean?"

"We have a nice fenced-in yard, we have a mudroom, there's plenty of space. It's perfect." Before I moved in with Jake, he lived in this house with two of his friends. I had heard all about how his friends tried to convince Jake to adopt a dog for two years, but he wouldn't budge. I wanted to introduce the concept lightly to assess where he was at with the whole dog topic.

"Yep! It's perfect for dog sitting," he said.

"Yeah . . ." I paused. "What do you think about us having our own dog?"

Jake chuckled. First his roommates and now his girlfriend. He could not escape the question. "It'd be really fun, but there'd be a lot to consider. They take so much work and they're expensive, and it might be more difficult for us to travel," he said. My heart sank. Jake noticed my sudden change in disposition. "I know we'll have a dog someday," he said. "We just moved in together, let's give it some time."

My feelings in that moment told me how ready I was to have a dog again. I did not at all expect to be upset by Jake's perfectly reasonable response. I didn't want to argue, but I also didn't want to end the conversation here. A few minutes later, I changed my approach.

"So, let's just say, hypothetically speaking about someday, what kind of dog could you see us with?" I asked.

"One that's active, but not too big. Ozzie and Truman feel too big for this house," he said. "But they fit perfectly with Mandy's house." The tall ceilings in Mandy's house really did suit them better.

I imagined us adopting a thirty- or forty-pound, playful dog from a shelter. There were so many dogs who needed a good home, and I wanted to help one of them. Both of my older sisters adopted their first dogs from shelters and had great experiences.

Jake brought our pancakes into the dining room.

"I have an idea. Let's look at the Nebraska Humane Society website to see what kinds of dogs they have," I said.

Jake laughed. "Christina, why would we do that now if we aren't looking for one?"

"Because it's fun. Why not look at pictures of dogs and see what we like? There's no harm in that." Without giving Jake a chance to respond, I ran upstairs to grab my laptop. Jake ate his breakfast, barely looking up at the screen until he caught a glimpse of an image of six or seven chocolate lab puppies cuddled together in a shopping cart.

"Look how cute the puppies are. Christina, look!" Jake grabbed my laptop and zoomed in on their faces. What was happening? Jake was supposed to be the practical one.

"Yes, they're very cute, but there's no way we're getting a puppy," I said.

"What do you mean? We'd be missing out on the best phase. Plus, how could you resist that cute little face?" Jake zoomed in even farther.

"Puppies take so much work."

Jake was only four years old when his family had brought Kirby home. I was older when we added Wrigley to our family. I remembered how much time it took to potty train her, how she could not be left alone for more than a couple of hours at a time, how my dad slept by her kennel every night for a couple of weeks, how we always had to keep an eye on her for months.

"I don't even know how we'd do that," I said. "Let's keep looking at the adult dogs." As we scrolled through the pages,

Jake continued pointing out every puppy and I continued staying focused on my mission.

Later that afternoon, Jake and I unpacked some more of my boxes upstairs. My parents were coming to stay with us the next week, so we wanted to finish setting up the house before they arrived. While we were working, we kept hearing *ding ding ding* from downstairs. I would walk down to our living room, see Ozzie sitting next to the bell staring at me, and I would try to take him outside. Instead of going outside, Ozzie walked back to the door and rang the bell again. After the second or third time of this confusing game, I noticed that the water bowl was empty. "Oh, you need water?" Ozzie licked his lips and ran to his bowl. He and Truman took turns taking long drinks.

Ringing their bell was just supposed to mean they wanted to go outside, but I started noticing that Ozzie seemed to ring it for all his needs. I wondered if he used the bell for everything because that was his only option. It intrigued me. I started imagining the potential for all Ozzie and Truman might have been trying to communicate this weekend. *What if they rang the bell to let us know they were hungry or needed water or wanted to play? What if they wanted us to pet them or to take them somewhere specific? If they only had one bell, how could we know what else they might have wanted?* But my mind quickly shifted back to the current task at hand: fitting all my clothes into the closet.

That night after Mandy picked up Ozzie and Truman to take them home, Jake and I sat in the sunroom in silence. My dad always talked about how once you live with a dog, you cannot go back to life without one. When you are used to a dog's unconditional love, playful spirit, and companionship, life seems too boring and incomplete without one by your side. I remember feeling this way within days of our family bringing Wrigley home. Even though I had lived ten years without a dog, I suddenly could not imagine life without one anymore. Did I really used to walk through the door without someone greeting me? Did I really used to play in the backyard by myself? What did we all do together before Wrigley? Since Jake grew up with an outdoor dog, this was the first time he really experienced a home feeling empty.

"So what do we do now?" Jake asked. "It feels so weird without them here."

"I know what we could do," I said, smiling. "We could look at more dogs."

CHAPTER 3

The Chocolate-Colored Puppy

The night that Ozzie and Truman left, Jake agreed to start looking around at dogs. I spent nearly every spare moment the next two days scrolling through online listings of adoptable dogs in Nebraska, Iowa, Illinois, and Kansas, and swiping through pet finder apps on my phone. Our perfect match could be anywhere. I did not want to keep ourselves from finding him or her by restricting the distance. We stopped our scrolling when we saw a one-year-old mutt named Colt from a shelter about an hour away. He was well trained but still young, and he smiled in every single picture.

Jake and I answered each question on the shelter's application with confidence and ease. It seemed like we had every criterion necessary for potential dog owners: solid financial situations, experience with dogs while growing up, a large fenced-in backyard, active lifestyles, plenty of space in our home, and Mandy's contact information to reference how we cared for Ozzie and Truman. The application probably could have been completed in twenty minutes, but Jake and I spent over an hour sitting on the love seat in our sunroom, savoring this vision of a life with a dog.

The last question was the only one to trip me up: "Under what circumstances would you consider relinquishing a pet?" The only situation I could think of was if we had a baby someday and the dog harmed him or her. But did I really need to express that? I figured it couldn't hurt to be as honest as possible. I jotted my response down, pushed submit, and headed up to bed. As I drifted off to sleep, I thought, *Tonight could be the start of our story of bringing our dog home.* I could not wait to hear back from the shelter and experience the rest of it.

When I woke up the next morning, I already had a response in my in-box. The adoption coordinator said everything looked great. She needed to officially meet with the rest of the adoption committee and would reach back out to schedule a time for us to visit over the weekend.

For the rest of the week, Colt was mentally mine. I had no doubt that we would bring him home. I called my parents to share the exciting plans, brainstormed with Jake about where to keep his bed and dishes, and talked about him nonstop to my friends. But on Friday night, I became nervous. The shelter still had not reached out again to schedule our visit. *Could he have already been adopted?* I wondered.

When my parents arrived the next day, it was still radio silence from the shelter, even after I reached out again. It wasn't until Sunday morning when the four of us were putting our coats on to leave for brunch that my phone screen lit up with a notification from Gmail. Mom, Dad, and Jake all huddled around me. I read it out loud:

Our committee members expressed concern over the possibility of relinquishment if a child is added to the home. One of the common reasons dogs are relinquished to shelters is that the family is unprepared to handle the dog's needs—training, time and attention, and socialization with the new family member. We are unable to further consider your application, but thank you for your interest.

I was crushed. This had to be a miscommunication. I wrote the shelter back immediately, clarifying what I meant. I would only *consider* relinquishment if the dog was outwardly aggressive and dangerous to a child. I wasn't expecting that to happen, I was just being honest.

Everything around us matched our moods the rest of the day. The sky poured rain, we waited for close to two hours to eat at the brunch restaurant, the shelter didn't respond back to me, and every other dog Jake and I showed the slightest interest in had already been adopted. This all felt like a lost cause.

Later that evening, my parents, Jake, and I stopped back at our house to change for our dinner reservations. I announced that I was officially done browsing pet finder apps for the weekend. I think I looked at every adoptable dog between the ages of one and four in the whole state of Nebraska. I wanted to enjoy the rest of the time with my parents here and focus on anything else besides the dog pursuit that was leading me to dead end after dead end. Maybe it was not the right time for us.

At one point, I was talking to Dad but he was oblivious, staring down at his phone.

"Dad?"

"You guys gotta see this," he said, grinning. He handed me his phone and walked away. Jake and I exchanged confused glances and looked down at the screen. It was a picture of four adorable puppies standing together, perched on the edge of a blue tub. Three of the four puppies had speckled and patched fur patterns, but one puppy in the middle had a smooth chocolate coat with a striking white chest. That one stuck out from the crowd.

"These are the cutest puppies I've ever seen," I said. "Mom, come look!"

Jake took the phone out of my hands. "Where are they? What is this?" he asked.

"Craigslist," my dad said. "Just searched for dogs in Omaha."

The listing read "Catahoula/heeler puppies. One male, three females." There were four pictures of the puppies and two pictures of the parents. The only other piece of information was a phone number—no price, no description of the dogs, or anything else.

I decided to call. With the way this weekend had played out, I expected there to be no answer, or to find out that all the puppies were adopted already. I paced across the living room, biting my nails until a woman answered on the fourth ring. It turned out that her Catahoula had puppies with her brother's blue heeler. The three female puppies were still available, and she could bring them to the neighborhood Hy-Vee parking lot after dinner to let us meet them. All the despair from the day washed away.

We spent the entire dinner at our favorite restaurant, Stirnella, staring at the pictures of the puppies, and googling "Catahoulas," "Blue Heelers" and "Catahoula Blue Heeler mixes." The Catahoula is the Louisiana state dog described as "intelligent, energetic, independent, inquisitive, loving, gentle."[1] Their coats are typically spotted, but one variation, the "red Catahoula," can have a brownish-red coat. From the picture on Craigslist, the mom looked like she might be a red Catahoula. Blue heelers, also known as Australian cattle dogs, are herding dogs described as "intelligent, active, alert, resourceful, protective, hard-working."[2] Even though Jake and I were so excited to meet the puppies, we prepared ourselves to act rationally. We had figured out how our lives would look with an adult dog, not with raising a puppy. We wanted to be responsible and figure out if this could actually work before making any decisions. Jake and I convinced ourselves that we would meet the puppies tonight, sleep on it if we liked them, then take tomorrow off work to prepare our house and bring one home. After dinner, the four of us hurried to the car, trying as best we could to stay dry in the March showers.

"Just so everyone is clear, we are *not* getting a puppy tonight. We're going to look and then decide," I said. "We're all on the same page, right?"

My dad laughed from the back seat. "What have I always said, Christina? Nobody ever ends up just looking at puppies." When I glanced in the rearview mirror, I caught my mom looking out the window, trying to cover her grin with her hand. She somehow knew something magical was about to happen.

"Here we are," I said. The giant red Hy-Vee sign on the front of the building lit up the dark, misty parking lot. I pulled into a spot in the back corner of the lot and waited. Within a few minutes, a car drove up and parked directly behind us. Jake squeezed my hand and smiled. The four of us watched out the window. A dark-haired woman came out of the car snuggling three puppies against her bright blue sweatshirt.

The woman handed the chocolate-colored puppy to my mom, a tan-and-white spotted puppy to me, and a puppy with speckled brown fur to Jake.

"Whoa, I can't keep her in my arms," my mom said. The chocolate puppy leaped from her into my chest. She licked my face over and over again.

I laughed. I looked down and locked eyes with this sweet, tiny creature. How could she possibly be so excited to be held by a complete stranger? It looked like she was dancing in celebration of landing on me. But maybe she felt the same way I did, like I was reconnecting with an old friend instead of meeting someone for the first time.

"Jake, watch out, she's coming for you." The chocolate puppy licked my face one more time, then leaped over into Jake's arms.

Jake looked down into her golden eyes. "You are the happiest little girl."

Jake and I instantly bonded with this little chocolate puppy. She bounced back and forth between us, her tail slapping into us as it wagged with joy. She did not care what the other puppies, my parents, or the breeder were doing. She only looked into our faces, and the three of us entered our own little world, only to be interrupted by the need to decide what we were going to do.

"I love her," I told Jake. "What do you think?"

"I love her, too. Let's stick with our plan. We can always get her tomorrow if we're feeling the same way."

I kissed the puppy's head and whispered in her ear that everything would be okay. I handed her back to the breeder. The puppy's eyes drooped when she realized we were leaving.

She extended her front paws out toward us. When the breeder walked back to the car, the puppy kept her eyes on us. My heart sank. "I can't look anymore!" I turned away and jumped in the car. Why did I set the stage that we weren't bringing home a puppy tonight? She was absolutely perfect, and we just let her go. "I think this is why it didn't work out with Colt," I told my parents and Jake. "She's the one."

Jake and I spent the car ride home gushing about the little brown puppy. I wished she was in my lap right now. We walked into our big, empty house and began panicking that someone else would meet her and take her home before we came back tomorrow. I was not going to last the night.

"I don't care how much time it takes to train her or how much work it is. I want her and I know we'll figure it out," I told Jake.

Jake called his parents to sound them out about adopting a dog. The conversation only confirmed the idea. When he was done, he said, "Let's go get her tonight!"

My parents cheered. I frantically texted the breeder asking if we could come back now. It was already after 9:00 so I wasn't sure if she would respond or wait for the morning. The four of us sat around the coffee table watching my phone. Suddenly, the screen lit up. It was an email from the shelter. Right when we decided we wanted a different dog, they wrote back to say they reconsidered my application and we could proceed with the adoption process if we wanted. But that didn't change anything for Jake or me. We were set on the little chocolate puppy.

Within minutes, the breeder responded that we could come pick her up. We all hopped back in the car. This time when we pulled into Hy-Vee, the rain had stopped, and the breeder was waiting outside the car holding our girl. As soon as she saw me, she wagged her tail and reached out to me. "See, I told you it would be okay," I whispered in her ear. "You're coming home with us now."

Jake and I cuddled the puppy together while Dad ran into the store to pick up a dog bed, food, and a few toys.

"Have you thought about names at all?" Mom asked.

"How about Stella?" Jake said.

"That's perfect," I said. "Stella girl, our little star. We love you so much."

We found our newest family member and I couldn't have been happier. She was all ours to raise, to learn about, and to love.

CHAPTER 4

Communication Is Everywhere

At work the next day, I sat on the floor, alternating between stacking blocks with the blond two-year-old I was evaluating and talking to her mom about language development. There are many language milestones that occur even before toddlers start saying words. Words are only one form of expressive language. Parents' anxiety can be alleviated when they know which skills their child already demonstrates, recognize all the other ways their child is communicating, and understand which milestones to look for next in their development. Even though this mom's daughter wasn't talking yet, I had already learned so much about her language skills in the past hour we spent together.

"We'll go over all this in much more detail when I score the assessment and write up the report, but I want you to know that I saw a lot of great things today," I said. "She explored these new toys, engaged in play with me, used eye contact and several gestures to communicate during multiple different activities. These are all really important skills for learning words, so we're on the right track in a lot of ways. Do you have any questions for me right now?"

The mom took a deep breath. "I guess I just want to know, when will she start talking?"

Within this first year of practice as a speech therapist, I quickly learned that was the number one question I would be asked when evaluating children with language delays. At first, it flustered me. I didn't know how to effectively articulate that it was impossible to give an exact time frame, but there were a lot of indicators in a child's prelinguistic skills of how developmentally close they were to saying words. I spent many hours in between sessions rehearsing responses. I wanted to be realistic yet encouraging. And I wanted to be serious about the need for intervention, yet optimistic about the child's potential.

"It's impossible to give an exact time estimate, but the great thing about language development is that it's a gradual process filled with milestones along the way," I said. "It's kind of like trying to predict exactly when a flower will bloom. If a stem hasn't sprouted from the ground yet, we know we probably won't see a flower in the near future. We know there are steps that need to happen before a flower would appear. But if there's a closed bud on top of a stem, we know just about everything is in place for the flower to blossom next. When we go over the assessment results, I'll show you exactly where your child is at in her development, and what steps we'll need to see before we can expect her to say words."

After I walked the mother-daughter pair to the front desk to schedule their next visit, I hurried to grab my keys from my desk. Yesterday, on our first day with Stella, Jake stayed home from work. We needed a day to figure out the logistics of how we would let her out every couple of hours with our work schedules. My last appointment of the day didn't end until 7:30 P.M., and Jake worked a typical eight-hour day. He would head into the office earlier now to be done sooner, and we both would come home on our different lunch breaks. Before we had Stella, I was always annoyed that my hours were later than everyone else's. Now, it felt like a gift that our different availabilities fit together to solve this Stella care puzzle.

When I arrived home and opened our bedroom door, Stella was sitting up perfectly straight in her kennel, waiting for me to let her out. "Stella girl, hi!" Her tail wagged frantically. She jumped up to lick my face repeatedly and flopped over on my lap. "Hi, girl. I missed you, too." I used one hand to rub Stella's belly, and the other to pat the towels in her kennel. *Phew*, I thought. They were all dry. I had been worried all morning about how Stella would react on our first day away.

"Yay, Stella, good girl! Let's go outside, come on." I carried Stella down the stairs. "Outside, outside, let's go outside!" I repeated important words for Stella so she would be able to recognize them and respond. When I used to ask our family dog, "Wrigley, want to go outside?" she would run to the back door if she needed to go to the bathroom or wanted to play. With enough repetitions, Stella should be able to associate "outside" with going out to our backyard, too. When Stella squatted down in the grass, I celebrated even more than I did after realizing her towels were dry. "Good girl, Stella!" I scratched behind her ears and kissed her forehead. "Come on, let's go inside now."

I loved watching eight-week-old Stella explore our home like an adventurer stepping onto uncharted land for the first time. She sniffed everywhere she walked, picked up and dropped each toy she passed, and darted her head toward every new sound. She was fascinated by all that surrounded her. Stella approached her dishes and pawed her water bowl. "Oh, you need water? Let's get more water, Stella." In only two days of living here, Stella learned what each dish was for. She even gestured by pawing her dish to let me know that she needed more water. Her communication was already so clear. "Thanks for showing me, girl. Here you go." Stella started drinking before I could even pull my hand away.

Nonverbal communication plays a crucial role in language development. Not only is it a significant milestone on the way to words, but research shows direct correlations between a child's gestural repertoire and subsequent expressive vocabulary.[3] This means children use gestures before they are able to say a word for the given concept. For example, a child who says "up" when he wants to be lifted likely reached his arms up to communicate that desire before he started saying the word *up*.

"Okay, Stella, it's time for another nap," I said. "Jake will be home in a couple hours to play, I promise." Stella trotted into her kennel sniffing to find the treat I tossed in. I closed the door and resumed the playlist of soothing music I made for her. I didn't want Stella to feel alone while we were gone. "Bye, Stella, love you." Stella watched me as I walked out the door. How could I leave that little face?

Back at work, I was reviewing a toddler language assessment I had completed when something struck me. *If Stella is already gesturing at eight weeks old, what other communication skills does she display that overlap with those of toddlers?* I took a closer look at the prelinguistic skills listed across the areas of "Interaction," "Pragmatics," "Play," "Language Comprehension," and "Language Expression." Turning through the pages I quickly realized, there were parallels in each domain. These are just a few examples of milestones that jumped out at me after being with Stella for only two days:[4]

- "Cries to get attention." Stella cried out to Jake and me, especially at night.
- "Turns head to a voice." When Jake and I talked to Stella, she looked back at us.
- "Anticipates feeding." When we walked to her food shelf, Stella stood by her dishes, waiting for her meal.
- "Maintains eye contact." Stella locked eyes with us while we were playing or talking to her.
- "Attempts to interact with an adult." Stella dropped her toy at our feet, whined, and barked to engage with us.
- "Responds to request to 'come here.'" Stella was learning how to "come" on command. If we patted our legs and crouched down to her level, she ran right to us.

- "Shows a desire to be with people." Stella followed us everywhere we went. If she was startled by a loud noise, she would hop in my lap.
- "Vocalizes to gain attention." Stella barked and whined to make us pay attention to her.
- "Searches for the speaker." If Jake or I started talking from the other room, she ran looking for us.
- "Plays fetching game with caregiver." Ironically, one of the first games young toddlers learn to play is running after a ball and bringing it back to their parent. Stella was learning to do the same thing. She chased her ball and brought it back about half the time.
- "Explores toys." Stella tried out every toy we had for her. She was learning which objects were hers, and which ones were not.
- "Gestures to request action." Stella rolls over to request a belly rub. She pawed at her water bowl to tell me to fill it up.

All these subtle behaviors are actually indicators of a child's language development. I saw all these, and more, in Stella. It made me wonder, *What is Stella's potential for learning and using language?*

After work that day I returned home to a frantic Stella, who was playing with Jake. Jake squeezed a pink plastic ball and bounced it across the living room. When she heard the high-pitched squeak, Stella abandoned a tennis shoe she had picked up and pounced on her toy.

"Good girl, Stella! Good girl playing with your toy," I said.

Stella squeaked her ball a couple of times, then dropped it when her eyes landed on our ivy plant on an end table. She jumped toward the plant, pawing at the leaves.

"Stella, no," I said, firmly. She stopped pawing the plant and looked at me.

"Good girl! Play with your toy." I rolled her ball right in front of her to distract her from the plant. She took the bait and ran through the living room with her ball in her mouth.

"Yes, good girl, Stella. Good Stella," I said.

Jake was impressed. I wanted to teach Stella what was hers and what wasn't by positively praising her every time she played with her toys. When she approached a plant, shoe, or pillow, I would say "no," redirect her to what she could play with instead, and show her how excited I was when she chose one of her toys. I thought back to the toddler language assessment. One of the comprehension skills is "responds to 'no' half the time."[5] After two days in our home, she was beginning to respond to "no." I started tallying in my head.

Later, Stella approached the staircase, and Jake and I followed right behind her. She put her right paw on the first stair, then pulled it back to the ground. She hovered her left paw over the first step for a couple more seconds then brought it right back down. Stella turned to look at me, then looked back at the staircase and whimpered.

I scooped her up and walked up the stairs, analyzing what Stella just did. She wanted me to see that she needed help going up the stairs, so she used her eye gaze to direct my attention to the steps. And she even combined that gesture with a vocalization, her whimpering. That's a skill called *joint attention*, a form of communication kids develop before they can say words. I write about it every day at work in my session notes.

Joint attention is a huge component of communication. It's when two people are focused on the same object or activity. One person draws attention to the object verbally or nonverbally, and the other person responds by looking at it. Anytime we tell someone, "hey, look at that" and point, we are engaging in joint attention. When toddlers are developing this shared attention ability, they first communicate with eye contact or other gestures, then they pair a gesture with a vocalization, then they say words to grab attention.[6] Stella just paired a gesture with a vocalization. That typically comes right before we hear first words.

For the rest of the night, I noticed Stella's joint attention in all kinds of situations. She did not only initiate it once on the staircase. She was constantly using eye contact and vocalizations to show me or Jake what she wanted, or how she wanted us to participate in her play. While sitting on the couch with us, Stella dropped her toy to the ground. She looked down at the toy and looked back at me, showing that she wanted me to pick it up.

"You want me to play?" I picked up Stella's toy from the ground and tossed it in the air for her. Stella bit it a couple of times, then dropped it right back down to the floor and looked up to me. She wanted to play the game she created again. It is like when a baby learns how to play "peekaboo" and repeatedly covers his eyes to start the routine again. After the third time, I started responding as if Stella were a child.

Instead of talking to Stella in sentences, I dropped down to using single words at a time. When children are learning language, it's important to spend time speaking to them at the level right above their current expressive capabilities. If a toddler is not talking yet, I say words like *want* when he reaches for a toy, *more* when I am about to activate the toy again, and *all done* when we are cleaning up. When a child starts saying single words, I increase to saying two-word phrases like "want toy," "more play," "all done play," and so on. With this strategy, I model the next level of communication without overwhelming the child with complex sentences to dissect.

Each time Stella dropped her toy and looked at me, I said "play" before picking it up and handing it to her. Stella and I were building a game together and connecting to each other. I watched and listened to her, and she watched and listened to me. If she dropped the toy and did not look at me or vocalize, I would not reach down for her toy. I waited to say "play" again until she communicated by looking down or whining. This way, I was adding the corresponding word to her nonverbal communication. By the end of our five-minute play session, Stella started wagging her tail every time I said "play." After hearing the word so many times and playing with me and her toy, she began to anticipate what would happen next. By pairing the single word *play* with her gestural communication and engaging in this game with her, Stella was starting to understand what the word *play* meant.

Communication is everywhere when you know how to look for it. On one of my very first days of graduate school, my AAC professors instructed us to write down observations while they showed our class a short video of a speech therapy session. When they collected our papers, I panicked. *What was I supposed to be looking for? What should I have observed?* I didn't know anything yet. Embarrassed, I turned my mostly blank paper over before passing it to the end of the row. On the last day of the semester, our professors played the exact same clip for us. Once again, we were to write down everything we noticed. This time, my pencil could barely keep up with my brain. I saw a rich communication exchange unfolding between this teenage boy and his speech therapist. I did not even realize we were watching the same video from the first class until our professors handed back our original papers.

One of my six bullet points from my first day of class was, "This boy uses a communication device to talk." On the last day, I filled the entire page with comments on his gestures, word choice, level of independence, vocalizations, different functions of communication, the clinician's reactions to his words, language facilitation strategies that worked or did not work, how much time he needed to process language spoken to him, when he became distracted, when he appeared focused, when demands appeared to be too easy, and when demands appeared to be too difficult. I stared down at my two lists in shock. There was so much packed into those five minutes. The video clip did not change. My awareness did.

The first couple days taking care of Stella felt like that last day of class in graduate school. I have played with plenty of puppies, but none since I became a speech-language pathologist. Stella was not the only puppy exhibiting all these human developmental milestones. My lens had changed since I learned this information and applied it every single day at work. Stella was already bursting with communication. If she were a child, I would expect to hear her first words soon. But, without the ability to develop verbal speech, what could come next for her?

My thoughts swirled. Dogs can understand words. Wrigley recognized all our names, *walk*, *outside*, *cheese*, *treat*, *kennel*, *play*, *toy*, and I am sure many others. Whenever she heard us say one of those words, she turned her head, wagged her tail, and ran to the appropriate location. Like Wrigley, most dogs wait around all day to hear favorite words like, *walk* or *treat*. Stella, a two-month-old puppy, was already exhibiting over half of the prelinguistic skills that nine-month-old babies demonstrate before they say first words. We have the technology now to say words in different ways, not only with verbal speech. The era of electronic communication devices began in the 1960s. A hospital volunteer noticed that the only way patients who were paralyzed could communicate was by ringing a bell. He created a typewriter that could be activated by inhaling and exhaling into a sip and puff gooseneck switch.[7] By the 1980s, several different large, portable voice output devices were made to help people of all ages with varying impairments talk if they could not rely on verbal speech. And in 2010, with the release of the Apple iPad, AAC language systems became much more widely available and affordable as apps that anyone can purchase. *What if Stella had access to a few different words?* I wondered. *Could she use an AAC device?*

Throughout the night, Jake and I alternated taking Stella outside when we woke to her cries. It was impossible to tell if she was whining because she was scared and wanted to be with us, or if she needed to go to the bathroom. She was trying to tell us something, I just did not know what.

I picked Stella up, "Come on, let's go outside," and carried her down the stairs to the backyard. My slippers crunched on the frosted grass. I stood shivering in the pitch black. "Come on, Stella, go potty."

Stella sat right against my leg and stared up at me. Even she looked confused as to why we were outside in the middle of the night.

"All right, guess you didn't have to go. Come on, back inside."

An hour later we woke to her whimpering again. "Let's see

if she'll stop in a few minutes," I said. "Maybe she'll go back to sleep." Right when I drifted off, Jake woke me up.

"Do you smell that?" he said. "I think she went to the bathroom." Jake hopped out of bed and turned on the lights. "Yep. Let's get some new towels for her."

The next morning at work, I dropped by my coworker and friend Grace's desk. She was also a first-year speech therapist and shared my enthusiasm for AAC. The two of us went to AAC conferences together, experimented with different devices in our spare time, and continuously worked toward all the children on our caseloads being able to express themselves.

"Why can't dogs use AAC?" I asked.

"Who says they can't?"

"I don't know, maybe no one. But don't you think dogs could say at least a few different words with a device? I was looking at these toddler language assessments and noticing how Stella does so many of the same things."

"Totally. I don't see why not. You should look it up, I'm sure there's something somewhere about it," she said.

I agreed. I thought there would be an abundance of research on dog AAC and the most effective practices. In recent years, multiple studies have confirmed that dogs understand words. In 2010, a retired professor, Dr. John Pilley, published a research study showing how he taught his border collie, Chaser, to learn the names of over one thousand different toys.[8] Dr. Pilley used information about how children learn the meanings of words to guide his teaching. He published another paper explaining how Chaser learned to understand sentences containing a prepositional object, verb, and a direct object.[9] Chaser did not just understand single words; she understood what they meant when combined together.

A study in 2016 proved that dogs do not just respond to the human's tone of voice like many have assumed. Dogs can comprehend the meaning of a word, and the intonation separately, like humans do.[10] One of the researchers from the study expressed that "dogs process both what we say and how we say it in a way which is amazingly similar to how human brains do."[11]

I figured with all these research conclusions, giving dogs a chance to say words would be the next logical step. There is so much research on effectiveness of different types of AAC, when to introduce it, why it works, language facilitation strategies to use, expected outcomes, everything for humans. I wanted to read what the research said for dogs and follow their conclusions.

We searched, "dog AAC research," "speech therapy for dogs," "dog communication device," "dog language use." The closest we could find were companies discussing communication devices that could potentially translate a dog's whines, barks, or gestures to words for their human to understand. But that was not what I was looking for. That would be like a device scanning our facial expressions and gestures and translating those into sentences. That would be much different from saying your own words and expressing your individual thoughts. I wanted to give Stella the opportunity to say the words she was hearing us say all day long and already responding to. If she can understand the words, she should be given a way to use them too.

"Darn," I said after clicking through four pages of results, coming up empty-handed. "I really thought there'd be something. I'm sure Stella could use a device if she had one."

"Let's keep brainstorming later," Grace said. "There's all kinds of AAC out there."

The next day when Grace and I had thirty minutes of open time that lined up, we revisited our dog AAC ideas. *What could Stella physically access?* and *What would be easy to test out?* were our two main questions. The icons on the tablets that Grace and I most frequently used with children would be too small for Stella's paw or nose. We briefly discussed the idea of an eye gaze system, which is a device calibrated to say the word for the icon that the user stares at for a few seconds. But eye gaze devices are very expensive and I did not even know what Stella's vision of a screen looked like.

Some children who have more significant motor limitations use a method called *switch scanning*, instead of directly tapping an icon with their finger. Switch scanning involves pushing a large button hooked up to activate the selected icon on a communication device. The user pushes the button to scan through the options on the screen, then double taps the button after landing on the desired word or pushes a second button to select it. The idea of these buttons seemed promising, if we could program them to say individual words.

"Wait, what about a BIGMack?" Grace said. With the capability to record anything into it, the BIGMack closely resembled a Staples "Easy" button. Individual BIGMacks cost more than $100, so I googled "recordable AAC buttons," in hopes of finding a more cost-effective solution. "Hey, look at those," Grace pointed at the screen.

I clicked on a link for Learning Resources "Recordable Answer Buzzers." They came in a package of four buttons, each a different bright color. "Wow, so much cheaper," I said. "I'm going to try them! What's the worst that happens, Stella can't use them? At least I would know."

TAKEAWAYS FOR TEACHING *YOUR* DOG

- **Observe how your dog already communicates.** Keep an eye out for your dog, whining, barking, pawing, wagging her tail, or using her eye gaze to direct your attention to something. Knowing when your dog is trying to tell you something will help you respond to her communication and determine which words to start teaching.
- **Respond as best you can to your dog's communication.** We are all more likely to communicate when we know we will be acknowledged and understood. Responding to your dog's verbal and nonverbal communication are equally as important as responding to words down the line. Build a solid foundation by observing your dog, learning her habits, and responding when she is trying to tell you something.
- **Pair words with your dog's communication.** Just like how I started saying "play" every time Stella looked down to her toy and looked at me, say a word that goes along with what you think your dog is trying to tell you or what your dog is doing. If you're not sure what to say, you can ask yourself, *What is happening right now?*

CHAPTER 5

Three Feet from Gold

One week after bringing Stella home, I tore open the box of Recordable Answer Buzzers that were delivered to my doorstep. Each button fit perfectly in the palm of my hand. The plastic black top curved into its wide base, a small red tab for recording was on its side, and the speaker and battery holder were on the bottom. I pressed the black cap down several times in a row. It was easy to push, which was important.

Stella weighed less than ten pounds at the time. She would grow and become much stronger over the next year, but I wanted to set her up for success from the very start. It is important to choose an AAC system that the user can physically access, or learn to access with practice, that will also support any future development. These buttons met that criteria. The force I had seen her use to paw at her toys and water bowl should be more than enough to push down the buzzers. When she reaches her full-grown adult size, they would not be too small for her paws either.

I was filled with the same anticipation and positive expectation I have right before trying a new AAC device with a client. It's a time that holds so much hope and potential for all that is to come. Even if a huge communication breakthrough does not happen immediately (which it normally doesn't), we are starting a path of discovery that can only help us all in the long run. It's like hopping in the car to begin a big cross-country road trip. You know you will not reach the other side of the country on the first day, and the trip will require hours of driving. But you will see signs that you are moving in the right direction along the way. You pass through different states, observe new landscapes out the window, and see landmarks in person that previously you had only ever read about. In the worst-case scenario, you realize you want to take a different route and adjust your GPS to navigate a new course. Even in that situation, you are still closer to your destination than you would have been by staying at home, wondering what the rest of the country looks like.

I found a couple of batteries. We could start with one button for now. The most immediate need we had for Stella's communication was a way for her to let us know when she needed to go to the bathroom. Now that she had been in our house for a week, Stella knew the door in the kitchen led to our backyard. She would not walk to the door on her own to let us know she had to go out yet, but she was starting to head that way when Jake or I would say "outside." Since Stella was starting to respond to "outside," I wanted to give her the opportunity to say it, too. I held down the recording button and said "outside" into the button's microphone.

I placed the green buzzer right next to the back door in the kitchen. The colors of the buttons wouldn't matter for Stella. Dogs have dichromatic vision, which means they do not see the same spectrum of colors that most humans see. Dogs' vision has been compared to a person who experiences red-green color blindness. At Stella's young age, when she showed any signs that she had to go to the bathroom, we had a very narrow time slot before she would go. There was no waiting for us to figure out what she needed. If she started sniffing the ground, she would go to the bathroom a few seconds later. Now, if she would be able to tell us "outside," she could be in the backyard within a couple of seconds. Keeping the button far away from the door might lead to more accidents on the way out.

I was ready to officially start exploring my big question: *What would happen if I implemented speech therapy language interventions with my puppy?* Our kitchen transformed into a treatment room in my mind. I sat down next to the door to be at Stella's eye level. Jake leaned against the counter, like most parents observing their child's speech therapy session for the first time. They are eager to watch, yet hesitant to participate right away.

Stella walked to our pantry, raising her head up to sniff the lowest shelf filled with cereal boxes, bread, and crackers. I crawled over, attempting to redirect Stella away from the food. "Stella, look!" I crawled back to the door and pointed at the button. She looked back at Jake, over to me, back to the pantry, straight ahead to the door. She might have glanced at the buzzer, but she definitely did not look at it intentionally. I slowed down my rate of speech, articulated each and every syllable, and paused in between every word I said.

"Outside," I said both verbally and with her button at the same time. "Outside. Let's go outside!" Stella stared ahead at the closed door, still unaware of the new addition on the floor. I pushed the button to hear "outside," once more and promptly opened the back door. "Come on, girl!" Stella followed me, trotting down the three steps that led to our yard.

"Yay! Stella outside. Outside!"

Jake joined us. "Is that normal?" he said. "It seemed like she wasn't paying attention."

"It's okay. It takes time for kids to pay attention to a device or watch what I'm doing. So I'm sure it'll take time for Stella, too. At the very least, she's still hearing the words I'm saying and making associations between what she hears and what happens next."

Two important parts of teaching words are modifying how I talk to clients and creating a language-enriching environment for them. If children have a language delay or disorder, that means they are not picking up on language concepts by hearing people talking naturally in their environment. They need focused intervention to help them develop specific concepts. The way I talked to Stella in the kitchen is a combination of two of the most effective, evidence-based speech therapy techniques: *aided language input* and *focused language stimulation*.

Aided language input means using the learner's AAC system to talk naturally while saying the same words verbally at the same time. Using both spoken language and the device simultaneously is shown to increase the learner's receptive *and* expressive language skills.[12] This simple strategy improves language outcomes for a couple of big reasons. First, the speaker tends to slow down their rate of speech as they go back and forth between saying a word verbally and with the device. A slower rate of speech gives the learner more time to process the words. Second, aided language input serves as a *model* for the user. The learner hears a word verbally, hears and sees another way to say that same word, and observes the word being used in appropriate contexts.

Focused language stimulation involves repeating target words that are relevant to what is happening at the time. In a single interaction or activity, a word should be repeated at least five times before moving on. When Stella and I were about to go to the backyard, I said "outside" five times between my verbal speech and using her button. When we were in the backyard, I said "outside" another two times. That means, during this thirty-second period of getting ready to open the door and going into the backyard, Stella heard "outside" seven times. If we take Stella outside six times per day, she would hear "outside" around forty-two times. If we took her outside ten times a day, which was more realistic during this potty-training phase, she would hear "outside" seventy times in one day, all during the appropriate contexts.

Simply changing the way we talk to children and how often we model words at the right time makes substantial differences in their language development. One research study published in the *Journal of Speech and Hearing Research* shows how dramatic differences can be between toddlers' language skills before and after receiving focused language stimulation. At the end of the four-month-long study, the group of toddlers in the experiment possessed 400 percent more words in their expressive vocabularies than when they began. They also produced 257 percent more utterances during a typical play session than they did before receiving therapy. These are huge differences. The results for the experimental group were significantly higher than the statistics from the control group of toddlers who did not receive intervention during those four months.[13]

Late-talking toddlers are much more likely to understand new words and say new words when focused language stimulation is a key part of their treatment plan. Also, AAC users are much more likely to learn how to use their device when the adults around them use aided language stimulation.[14] I did not know if Stella would ever learn to say words, but I wanted to give her the best shot I could. If anything would work, the most effective teaching strategies for kids would probably be a great place to start.

At the end of the week, Stella had not shown any progress with "outside." We took Stella on her first walk around the neighborhood and watched her try climbing her way out of her first bath, but Stella did not show any awareness of the button next to the door. I was used to this happening with kids. Sometimes it took weeks for a child to start really noticing the device I was using. Knowing this, I had motivation to keep going when there were truly no signs of progress after several days. If I had not introduced an AAC system to a child or taught language before, I am sure I would have thrown in the towel pretty early on. But Stella's lack of progress actually propelled me to buy more batteries and program two more words for her. (I would have done all four, but the fourth buzzer was broken.) I was excited to see if using three buttons instead of one would make a difference in Stella's learning.

I had encountered a similar situation when a three-year-old boy with autism transferred over to my caseload. He had never said a verbal word. He only ever squealed when he was excited and grunted when he was frustrated. The previous clinician wrote that she trialed a single button programmed to say "more" for five months. He showed no progress, therefore she questioned his ability to learn any AAC at all.

When I started working with him, I incorporated a device with a robust language system. It had the capacity for him to eventually say thousands of words. Within one month of the change, he used his tablet to say single words such as *up*, when he wanted to go upstairs, *like* when he was excited about a toy I brought, *on* while his occupational therapist helped put his shoes on, *get* when he wanted a toy that he could not reach, and *more* when he wanted me to start his favorite toy again. Instead of seeing his device in use only when it was appropriate to say "more," he saw me using his device constantly for a variety of words. Greater exposure leads to greater learning.

I wanted to pick words that would be meaningful to Stella. From observing and interacting with her for just a couple of weeks, it was very clear that Stella loved to play. She was often initiating some type of game with us. Whether it was chasing Jake around the dining room table, dropping her ball off the couch for me to pick up, or running after a toy, she loved having fun with Jake and me. I recorded "play" into the orange button and set it down in the living room next to her basket of toys. A verb like "play" leads to more opportunities for learning than nouns like "ball" or "toy" do. Stella could say "play" to request any type of game rather than a single one. That could come later.

The next concept that I thought would be helpful was "water." I tried my best to notice when Stella's water bowl was empty, but sometimes I would walk past it, realize it was empty, and fill it up. I hoped it had not been like that for too long. If Stella had a way to say "water," I would know right away when she needed more. I could also hear her request more water from another room instead of needing to see her pawing at her bowl.

Stella was already very aware of how I filled up her water dish. She would nudge her dish, keep her eyes glued to me as I carried it to the sink, and stand right where it would land. I programmed the magenta button to say "water," and set it right next to her bowl.

Since children typically start saying words that they already have gestures for, "play" and "water" made sense for Stella. She used gestures to communicate about both those concepts by pawing at her empty water dish, nudging her toys and looking up to me, and using her eye gaze to direct my attention toward a specific toy that she wanted me to pick up. I could not wait to find out if Stella would demonstrate the same gesture to word progression that occurs with kids.

Now that we incorporated three buttons instead of one, Stella definitely received more exposure to language and her AAC in the following days. Anytime we walked by Stella's dishes and saw her drinking water, we would say "water" and push her button on the way out of the room. Jake would chase Stella around the dining room table, then would stop to model "play" before he started another lap. I would hold Stella's toy basket and push "play" every time I pulled a new toy out of it. Before taking Stella on a walk, we modeled "outside." My friend Grace would come over for lunch, and we alternated between eating and modeling "play" or "outside" for Stella. Stella was now hearing the same few words from multiple people, in several contexts each day.

Celebrating small victories is an important part of being a speech therapist. Significant changes in communication take time to see. There are always little steps on the way to a new milestone that show us we are heading in the right direction. Celebrating those successes gives the child encouragement to keep trying, and his treatment team and parents inspiration to continue teaching. About a week and a half after introducing the buttons to Stella, we witnessed our first small victory to celebrate. I was grabbing a snack from the fridge, when I turned to see Stella standing by the back door, looking down at the "outside" buzzer. This was completely unprompted. I was not in the middle of modeling it, pointing to it, or saying "outside." She gazed down at the button for a whole five seconds. I closed the fridge and squatted down next to her.

"Yes! Good girl, Stella." I petted her head. Stella wagged her tail and licked my face. She got so excited when I told her she was doing a good job. "Outside," I said verbally and modeled with her button. "Come outside." This time, Stella watched me closely when I pushed her button.

There was no way for me to know Stella's intention as she stood in front of the buzzer and looking down at it. She might have wanted to get my attention to go outside, or she might have simply been curious about the object on the floor. Either way, I did not care. Providing positive praise showed Stella that I enjoyed what she was doing, which would hopefully lead her to continue exploring the buttons in the future.

When we came back inside, Stella sprinted through the house and dove into her toy basket. She pulled out her favorite yellow squeaky toy, tossed it in the air, then dropped it right at my feet.

"Play!" I modeled with her button.

Stella cocked her head to the side and looked down at the buzzer. That was the second time in a row that she showed any type of awareness of the buttons.

Instead of pushing Stella to try to do more in those moments, I praised her for showing interest in the buzzers and modeled use of them again. For almost two weeks, Stella had not appeared to notice our modeling, or even the presence of the buttons. Before now, she explored every other object in her space except for these. I hoped this was only the beginning.

Unfortunately, at the end of two weeks of modeling each word extensively at nearly every applicable time, the most Stella had done was look at two of the buttons. She had not looked at them again and did not demonstrate any other signs of progress. *How long should I keep trying this idea?* I wondered. We know all kids can talk in some way, no matter how long it takes to get there. There is no question about whether to give up or not. The answer is always to keep going. We do not know this about dogs, though. *Is this something that will take months? Will the results be worth the time I am putting in? Or is it something I will keep trying and get nowhere?*

Maybe this is why there was no research on dogs using AAC. Maybe the similarities between dogs' and humans' language capabilities stop here, at understanding words and communicating concepts with gestures. Over the next few days, I did not model words as much as I had in the beginning. I said "play" or "outside" or "water" once, and then pressed the corresponding button. But I did not provide as many repetitions, nor make sure I modeled language at every appropriate time. Sometimes I would rush Stella out the door without pressing "outside" first. Or I would fill her water bowl up in a hurry without pausing to model "water." At this point, I stopped expecting to see progress and decided it would be a pleasant surprise if anything more would happen. I did not completely quit my pursuit, but it was no longer a top priority for me. *If she hits a button, great. If not, no big deal*, was my official stance.

There's a famous story written in Napoleon Hill's *Think and Grow Rich* about a man named Mr. Darby who dug for gold during the gold rush. He found a vein of gold close to the surface level, went back home to raise the money he needed for the machinery to extract it, then returned to the mine. He extracted the gold he could see, almost enough to pay back the people who lent him money for the machinery. According to his calculations, there was supposed to be much more gold farther beneath the surface. He kept digging and drilling, digging and drilling, but there was no more gold to be found. With no further evidence of the precious metal, Mr. Darby quit his pursuit. He sold his machinery and went back home. A man bought Mr. Darby's equipment and dug three feet farther. He found an entire vein of gold, which resulted in millions of dollars. Mr. Darby had quit, just three feet from a life-changing find.[15]

Hill pointed out, "Failure is a trickster with a keen sense of irony and cunning. It takes great delight in tripping one when success is almost within reach."[16]

A few days later, at the start of week three with Stella's buttons, I had not modeled anything all day long. Stella sat on my lap on the couch, then jumped off and trotted into the kitchen.

I followed her to the back door where she stopped. Stella looked down at the buzzer, then up at me, and back down to her "outside" button. This time, there was no questioning her eye gaze. She was gesturing for me to push it, just like the time when she gestured for me to help her go up the stairs. I immediately pushed "outside" with my foot and ran into the backyard, praising her the whole way. She went to the bathroom as soon as her paws touched the grass.

When some kids are learning to use their AAC devices, they often grab my hand to push an icon for them. In my experience, this has always come shortly before a child uses his own finger to say a word, when he still needs a little more support before getting there on his own. With one meaningful look from Stella, my enthusiasm was back in full swing. Stella knew that I had been pushing the "outside" button before going outside. When she had to go to the bathroom, she directed my attention to her button to let me know what she wanted.

I came home from work the next day to find Jake smiling in the kitchen. "You were right, she's really aware of the buttons now," he said.

"What is she doing? What did I miss?"

"Give her a minute and you'll see for yourself."

I sat down on the kitchen floor and let Stella crawl all over my lap. This was my favorite part about coming home from work at the end of the day. "Hi girl, I'm so happy to see you." I kissed Stella's forehead in between her floppy ears and stood up. Stella walked around the kitchen, sniffing under cabinets and after a few minutes, she marched up to her "outside" buzzer. She stared down at it. When nothing happened, Stella paced past it three times. She stood in front of it again and lowered her front legs to the ground. She barked at the button and looked up at the doorknob.

"She's already done that three or four times tonight. Whenever I push it, she wags her tail and watches me until I open the door."

Stella combined multiple gestures with a vocalization. That is even more impressive than putting one single gesture and one vocalization together, which is what happens with kids right before a word. It reminded me of watching someone play charades. They come up with every possible gesture for a word without actually saying it. Maybe we were just three feet from gold. Maybe this really was going somewhere after all.

TAKEAWAYS FOR TEACHING *YOUR* DOG

- **Narrate what's happening in your dog's environment by saying single words and short phrases.** Talk about what your dog is doing, what you're doing, or what is about to happen.
- **Be repetitive.** Aim to say a word at least five to ten times before moving on to something else. The more often your dog hears a word in the right context, the faster she'll be able to learn its meaning.
- **Program words into your device that are frequently occurring, and relevant to your dog's daily life.** Remember, a more general word like *play* will be more valuable and frequently occurring than the name of one specific toy.
- **Turn your dog's typical activities into teachable moments.** Before you take your dog outside, go for a walk, feed her, play with her, replenish water, or give her belly rubs, take a moment to say what you are doing multiple times.
- **Use verbal speech *and* your dog's buttons to talk to your dog.** Every time you use your dog's button to say a word, you are modeling appropriate use of it for her.
- **Teach more than one word to start.** Greater exposure leads to greater learning.
- **Be on the lookout for small changes in how your dog interacts with the buttons, responds to you, and communicates.** Your dog probably will not learn to functionally communicate with AAC overnight. That is completely normal for both humans and canines. Look for small victories along the way, such as your dog looking at the buttons, watching you use the buttons, standing next to the buttons, or barking at them. Give verbal praise anytime you see her exploring them.

CHAPTER 6

Now We're Talkin'

Stella had already changed so much in this first month since we brought her home. She grew a few inches taller, learned how to climb up and down the stairs, started sleeping through the night without Jake lying next to her kennel, made a couple of puppy friends, and began forming different relationships with each of our friends. Stella was so interested in people. She greeted everyone who came to our house with a warm tail wag and body wiggle followed by falling over in their laps, hoping for belly rubs. She watched all of us so closely, fascinated by our every movement. It entertained Jake, our friends, and me alike to see how Stella preferred to interact with each of us in a unique way.

Stella turned absolutely crazy whenever our high-energy, playful friend, Jenna, came over to visit. Stella would run around in circles at Jenna's feet, baiting her to chase her by lunging and sprinting away. When Jenna sat down on the ground, Stella climbed all over her to attack her face with kisses. But every time our calm, gentle friend, Alex, came over, Stella snuggled in her arms, looking as content as could be. When Jenna and Alex visited at the same time, we all cracked up watching Stella switch back and forth between the frantic and sweet sides of her personality. She would ever so gently lick Alex's hand and sit right by her side, then would leap over to Jenna, bark, and run to her toy basket. Stella's worlds were colliding. Watching her navigate this situation reminded me of parties I had with friends from separate eras of my life all gathered in the same place; different aspects of my personality would come out, depending on who I was interacting with at the time.

Wrigley was the same way. She had been excellent at reading people and adjusting her energy levels to match those around her. When I was a scrawny eleven-year-old, my parents were reluctant to let me hold Wrigley's leash on walks. She was strong and energetic, and notoriously pulled hard. She could have easily knocked me over with one bolt toward a squirrel. Much to everyone's surprise, the complete opposite happened. Wrigley slowed down and walked right by my side the entire time I held her leash. She knew she had to be gentler with me. Similarly, one weekend Wrigley stayed at my grandparents' house while we were out of town. My grandpa left to go on a fishing trip, leaving my grandma alone with Wrigley for the night. Wrigley had stayed there plenty of times and always slept on her own bed in the living room. But that night before falling asleep, Wrigley grabbed her bed with her mouth and dragged it all the way to the doorway of my grandma's bedroom. She stayed right there, guarding my grandma, for the whole night. Our entire family fawned over Wrigley's awareness of her environment and independent decisions to adjust her behavior.

Over our first four weeks with Stella, her personality unfolded more each day. She loved being near Jake and me but had no shortage of intrinsic curiosity or independence. She was well-behaved and intelligent, but not blindly obedient to what Jake or I wanted. She was social with strangers, yet wary of many new objects or sounds. I began to realize that it was hard to describe Stella with only one or two adjectives. But I could not summarize any of my friends' or family members' entire personalities with only a couple of words either. Stella is complex, just like we all are. When friends or family members who had not met Stella asked what she was like, the best single word I could use to describe her in a nutshell was "spunky."

Now after seeing Stella's newfound awareness of the buttons, I returned to modeling language for her every day, in nearly every natural opportunity. She no longer appeared absent-minded when I pushed a button and said the corresponding word. She watched my foot or hand intently as it activated the buzzer. She wagged her tail and looked toward the door, water bowl, or a toy depending on which word I said. Stella's wheels were turning. I felt like I could see her connecting the dots through her head turns, facial expressions, and tail wags. I stopped wondering if it would take months for Stella to use a button, or if she would ever be capable of saying a word. Instead, I started acting as though this milestone was bound to happen. It was the next logical step in her development.

While Jake and I ate dinner later that week, Stella lay on the floor, chewing on her rope toy in the living room. We sat facing her to make sure she stayed out of trouble. Suddenly, Stella dropped her toy and stood up. She started walking toward the kitchen. She paused when she reached the dining room table to make eye contact with us. As soon as she grabbed our attention, she continued on her path. She looked like a little kid making sure she had her parents' full attention before trying something new. Jake and I leaned over the table to watch her through the doorway. Stella stopped in front of the "outside" button. She stared down, like a child standing on the edge of the diving board mustering up the courage to jump into the pool. Stella lifted her right paw. I held my breath. She swatted. And she completely missed the buzzer by at least six inches. Jake and I exchanged glances and hopped out of our chairs.

"Good girl, Stella," I said on my way into the kitchen. "You can do it! Good girl."

Stella wagged her tail and lifted her paw again. She swatted three times in a row, again missing the button by a half of a foot with each attempt. "Yes, you're so close. It's right here." I pointed at the button, hoping one small cue to show her where the button was would do the trick.

In AAC therapy, best practice is to use a "least-to-most" cueing system.[17] Researchers developed a hierarchy that shows cues ranging from the most naturalistic to the most prompted. We begin with the most naturalistic cues such as providing an increased wait time of at least ten seconds, then move down to pointing at the device, or tapping the device. If the user still does not respond, then we model the word again and retarget the vocabulary. I never recommend grabbing a child's finger to say a word for him or grabbing a dog's paw to push a button. That would be the equivalent of pulling words out of a person's mouth, which is impossible. AAC users should be given the same amount of control over their own words.

Starting with the most natural ways to encourage communication and gradually increasing the level of assistance when needed is more beneficial than giving too much help right away and trying to reduce your involvement. The latter often causes "prompt dependency," a reliance on being told when to say a specific word instead of being able to communicate independently. The former builds a solid foundation for functional, independent language use.

Stella looked up at me and back to Jake. *Should I take her outside or give her a little more time to try again?* I asked myself. She was so close. I did not want to rush her out the door and cut this learning opportunity short. But I also wanted to reinforce how close she was to activating her button. Teaching language is a dance. It is a constant search for rhythm and balance between giving the right amount of processing time and reacting quickly, between praising progress and pushing to reach the next step, between helping too much and not supporting quite enough. There is not one exact formula to use in every situation. Even when you know all the tools and strategies to pick from, deciding which ones to try in each moment is an art that only improves with practice and reflection.

"Right here, Stella, right here," I said. I tapped the top of the button. Stella inched over next to me and squatted her hind legs down. A puddle formed right beside the button. Jake swooped in to pick Stella up and run her to the backyard.

"Well, that was 100 percent our fault, not Stella's," I said and sighed. *Why did I have to push for more?*

Even though Stella's attempt at saying "outside" ended with

a swing and a miss and her going to the bathroom on the floor, a lot of good came from that situation. Stella showed us that she was not only aware of the buttons' existence, she also was at least starting to understand their purposes. When Stella had to go to the bathroom, she let us know by walking to the button and trying to hit it. She demonstrated clear communicative intent. Seeing Stella attempt to say "outside" in such an appropriate context on her own was a huge step. I did not care that our twelve-week-old puppy had terrible aim. I was thrilled that when she had to go to the bathroom, she knew what to do.

"Christina, Christina, wake up. Stella said her first word!" Jake shook my shoulder the next night. I was tucked under the covers.

"What?" I reached over to my nightstand to turn the lamp on. It was 11:15 P.M. I must have only been asleep for half an hour or so.

Jake was giddy. He shoved his phone screen in front of my face. "Here, I wrote it all down so I could tell you everything," he said. "Wait, no, actually I want to read it to you." He snatched his phone out of my hands.

I couldn't help but smile as I sat up in our bed and listened as Jake told me the greatest bedtime story I had heard in my adult life. After he caught his breath, he began:

Stella was awake at 11 P.M. 4/30/18. She had lots of energy and trotted over to the kitchen door. She sat by it, so I stood and waited, and stared at her. She sat very patiently and kept looking up and down from the door to me then back. After probably 30 seconds of this (not whining or barking) she looked at me, got a little antsy, and then looked down at the button. She raised up her paw and hit the button! I was so shocked I didn't know what to say. She immediately hit it again and looked up at me expectantly. I opened the door, praising her all the way, and she went straight outside and peed! We stayed out for 5 minutes and she also pooped! Then we went back inside. We were inside for no more than two minutes, and she went back over to the door again and barked, and hit the button with her paw! So I took her outside and she SPRINTED around the backyard. She had so much energy. She is making so much progress and I couldn't be prouder of her!

I hugged Jake. That night, I was equally proud of him as I was of Stella. Jake did absolutely everything right, and he had no experience to guide him. This was all brand-new for him. It was late at night, and Jake could have just rushed Stella outside to go to the bathroom one last time, then put her to bed. Instead, he was incredibly patient. He gave Stella the time and chance to speak up for herself. He reacted to her word perfectly by showing excitement, going outside, and having fun with Stella. Part of me wished I would have stayed awake for another thirty minutes to see this happen in person. Mostly, though, I was overjoyed that Jake could experience the awe of watching another being develop a new skill right in front of your eyes and knowing that you helped them. There are few better feelings than that one.

The next night I came home from work, determined to see Stella say "outside" for myself. According to Jake, she had not said it again since last night. For about half an hour after dinner, I repeatedly modeled "outside" and took Stella out to play in the yard. Stella kept looking up at me with a head tilt and confused eyes every time I said "outside" again. We never went out this many times in a row.

At work it always drove me nuts to see parents trying so hard to make their child say something when they wanted him to, only to prove a point. I would gently say, "It's okay. I believe you. I'm sure he'll say it when it's meaningful to him," and then try to redirect the conversation. In this moment, I realized I was spiraling into exactly what I did not want. Stella would say "outside" again when it was meaningful to her. Communication is always a choice.

I joined Jake on the couch in the living room, distracting myself with an episode of *The Office*. Stella played with a plush raccoon then slipped out of the living room and into the kitchen. I did not notice she was gone until I heard a single high-pitched bark, the one Stella uses to call us over when she wants our attention. When I walked into the kitchen, Stella paced back and forth in front of her button. She looked back at me, then down at the buzzer. I paused for five or ten seconds to see if she would say "outside" on her own. Stella walked over to me, then back to the button.

"Outside?" I asked.

Stella barked.

I walked to the button and lightly tapped the top with my toe, cueing her on what she could do next. Stella's eyes met mine before she marched up in front of the buzzer, lifted her paw, and pushed down to say, "outside."

"Yes, let's go outside." I opened the door and called out to Jake before joining Stella in the backyard. "She did it again!"

Stella promptly went to the bathroom in the grass. It was just as I thought, she waited until it was meaningful to her to say "outside." That was so much more impressive and fulfilling than her saying it only to appease me. Jake and I celebrated another successful communication event by chasing her around the patio table. After a few laps, Stella noticed I had left the door open and pranced back into the kitchen. She turned around to meet my eyes again as I followed her. She said "outside" twice more in a row. She hopped back down the staircase to keep playing with Jake in the yard.

She was discovering her power. "This is incredible," I said.

I was absolutely thrilled that Stella had now said "outside" five total times before needing to go to the bathroom or wanting to run around and play. It amazed me how intentional her communication was, even from this starting point. She did not push a button and walk away. She expressed her desire to go outside first through her bark, eye contact, and pacing next to the button. Then she advanced from using gestures to a word right in front of my eyes. Saying "outside" was the cherry on top of an already rich communication sundae.

For a month, Stella had listened to and watched us say "outside." Through our modeling, she learned what always happened immediately after we said "outside." That is the power of providing naturalistic intervention and modeling words at their appropriate times. Knowing a word and knowing *how to use* a word are completely different skills. I was encouraged to see that so far, Stella said "outside" during contexts that made sense. These observations were still so premature, but I was curious to see what would happen next. *Will this skill last? Or will she become less interested after figuring it out?* I wondered. Would Stella use her "outside" button to communicate every want or need, or would she differentiate its meaning from other words? Would she start saying her other words soon, or would it take another month?

To my surprise, I started learning the answers to a few of these questions much quicker than I expected. Stella, Jake, and I returned to our living room, happy and out of breath from chasing one another in the backyard. Stella, with her bottomless energy supply, shoved her nose into her toy bin. I modeled "play" a few times while she searched for the toy she wanted.

Stella pulled out a ball and froze. She looked up to me on the couch, then looked back down, still clenching the ball in her mouth. With one confident swoop, she swatted her paw to push the "play" button.

"Was that you?" Jake asked.

"Nope, that was all her," I smiled. "Play? Yes, let's play. Good girl!" I popped up from the couch and bounced Stella's ball across the room for her. A couple minutes later, Stella said "play" again after she dropped her ball and picked up her rope toy.

After weeks spent talking to Stella, modeling words, and observing her communication patterns, that night, for the first time, she and I spoke the same language.

Stella needed a language input stage, hearing and seeing words in use, like all children require, before she could reach the language output phase. Now that we made it here, I was hooked. I opened my laptop and typed the details of her communication events. I wanted to remember everything about this night, the first time my puppy said a word to me.

A couple of days after Stella said her first two words, I stood in the kitchen, scarfing down my toast with jam. I was running late for work after I had stayed outside with Stella for triple the amount of our usual time, encouraging her to go to the bathroom in the rain. I knew if she didn't go then, I would be cleaning up an accident in the house. While I threw my plate in the dishwasher, I could hear the familiar slurping of Stella taking a long drink of water.

Suddenly, the sound stopped. She looked at the magenta button to the right of her water bowl. She then glanced at me, then back at the button. I peered over to see the bowl was now empty. After a few more seconds of hanging on to the silence, I tiptoed over and crouched down next to the "water" button. Positioning myself in this way was a gentle cue, meant to remind her of what she could do. Subtle prompts are like giving hints. It's like when you are trying to remember the capital of Montana. You're thinking and thinking, then when someone says, "It stars with H," you immediately say, "Helena." You did not quite come up with it independently, you needed a little hint to remember.

Stella understood my hint. She lifted her paw, and swatted it down to say "water" for the first time when her bowl was empty. I filled up her dish and Stella drank some more. Stella had now officially used all three of her buttons. I left bursting with pride.

At work, I showed my coworkers the videos of little Stella using three different AAC buttons to say "outside," "play," and "water." Grace was elated about the news. She helped me find these buttons, accompanied me on nearly every lunchtime trip to let Stella out, watched Stella's progressions with me in person, and brainstormed with me several times.

I thought everyone else would be excited, or at least interested in the topic too. I envisioned these videos of my adorable puppy communicating so intently with AAC inspiring other therapists to implement communication devices more and to be proud of how powerful speech therapy is.

Aside from a smile or a "cute" or "aww," people were mostly confused or uninterested. One did ask how I got her to push the buttons, and another simply said "cool trick" before swiveling her chair back to her computer.

In retrospect, my coworkers' reactions should have made a lot of sense to me. Unfortunately, it is common for speech-language pathologists to have little to no experience with AAC. That was true for this clinic. Fortunately, I attended a graduate school with an excellent AAC course. Not all graduate programs have comprehensive classes on augmentative and alternative communication. In fact, it's more common for the latter, meaning that speech therapists are left to figure out unfamiliar technology on the job.

Grace and I had spent the last nine months desperately trying to spread our AAC knowledge and enthusiasm to our coworkers. Together, we navigated the complicated process of submitting requests for AAC devices to be funded by insurance. We invited representatives from device companies to come speak to our staff at the clinic. We came into work on our days off to write an entire guidebook and create resources that explained how to use AAC, step by step. We even conducted AAC evaluations for other therapists' clients, wrote up the lengthy reports, handled all the paperwork, all while still treating the children from our full-time caseloads. We were fueled simply by the hope of more kids learning to talk and more therapists becoming reinvigorated by how much of a difference they could make. A few supervisors applauded our work ethic, but unfortunately it seemed like no matter what we did, we could not inspire anyone else to support AAC.

I did not care if anyone else saw the potential I saw yet or understood how exciting this could be for the field of speech therapy. I knew in my heart and in my head that I was witnessing something fascinating with Stella. If anything, my coworkers' confusion showed me how far our field still needed to come. I hoped someday I could be part of a movement toward more understanding and acceptance of AAC for all who need it, even dogs.

TAKEAWAYS FOR TEACHING *YOUR* DOG

- **Expect a language input phase.** Your dog will need to hear words to learn words and will need to see her AAC in use to learn how to use it. Expect that this will take time.
- **Model words when your dog communicates.** When you see your dog gesturing, whining, or barking, model the word for what you think she is trying to communicate. Pairing a word to her communication is powerful.
- **Use naturalistic cues to draw attention to buttons.** If a long pause doesn't work, you can stand by the button, point at it, or tap it. Pause again after giving a cue. If your dog still doesn't explore the buttons, go ahead and model the word and carry on with your activity.
- **Model during natural contexts.** Model words when they are relevant to what is naturally happening.
- **Respond to communication.** If your dog says a word, respond appropriately! In the beginning, try to honor your dog's communication as much as you can. If your dog is not using words, continue responding to all her other forms of communication. Do not withhold food, water, playtime, trips outside, or anything from your dog. Just create a minute or two of opportunity for your dog to try using a word.

CHAPTER 7

Independence

For the next few weeks after Stella said her first words, I rode a continuous, natural high. Summer weather arrived in Omaha, Stella said at least one word every day with less and less cueing needed, and I received my letter from the American Speech and Hearing Association (ASHA) granting me my Certificate of Clinical Competence. After four years of undergrad, two years of graduate school, and a yearlong clinical fellowship, I was finally a fully licensed speech-language pathologist, able to practice without a supervisor observing my sessions and reviewing my documentation. I had dreamed of the day when I would be a full-blown clinician, a licensed health professional, and it had finally arrived. I was proud of the career I had chosen, proud of the work I was doing, and proud of where I could go next. I also felt free to work wherever I wanted, and free to provide my professional opinion without someone signing off on it first. There was nothing more to do to prove that I knew what I was doing. With adding just three letters next to my name, my entire mind-set shifted. I sent more emails to kids' school therapists sharing my observations and recommendations, signed up for an AAC continuing education course to attend next month with Grace, and started thinking about where I wanted to work next. I could stay at my current clinic, but the world and all its job possibilities appeared so vast to me.

Stella still frequently walked back and forth past a button, barked at it, or stared down at it. When I saw these gestures, I would sometimes say a verbal prompt, like "What do you want?" Then I stayed silent for at least fifteen seconds to give Stella a chance to respond. Research shows that when communication partners pause for ten to forty-five seconds, AAC users are more likely to respond using their devices.[18] Giving a longer wait time cues the learner that it is their turn to talk. It also gives them more time to process what is happening then choose what to do next.

I stayed still. Stella stood with her back toward me, looking ahead to her button and her dishes. I counted in my head to keep myself from intervening too quickly. "12, 13, 14, 15, 16 . . ."

"Water," Stella said. She licked her lips and wagged her tail immediately, excited for me to replenish her bowl.

This was the most powerful cue we could give Stella to use one of her words now, simply waiting long enough to give her a real chance to say it. It is so natural for adults to want to jump in and help immediately, or to take over to accomplish something quickly. We have to become comfortable with silence and patience before we can expect to see real, significant progress.

One morning, while I was upstairs brushing my teeth, my thoughts wandered about the workday that lay ahead of me. I was mentally picking out the toys I would grab off the shelves when I arrived at the clinic, when my planning came to a sudden halt.

"Outside," I heard from downstairs.

Jake had already left for the day. So hearing "outside" could only mean one thing. Stella said a word completely independently for the first time. I ran down the steps, toothbrush in mouth, to find Stella standing next to her button, staring at the back door.

"Go outside? Yes! Good girl, Stella."

Stella wagged her tail and jumped around in a circle. As soon as I opened the door, she bolted outside, chasing a squirrel sitting on top of our wooden fence.

This was the first time Stella had ever said a word totally by herself, without any cues or modeling right beforehand. It was a huge milestone to celebrate. Before this morning, she needed at least one or two small prompts. This is completely common and natural for all kids, whether they are learning to talk with a communication device or verbal speech. Children need support before they can do something new entirely on their own. It's like when babies are learning to walk. They need to hold on to furniture or their parents before they can take several steps freely.

It was helpful for both Stella and me that I could be upstairs and know exactly what she wanted right when she wanted it. If she had barked, and come up to find me, I would not have known why. If she were standing quietly by the door, I would not have been able to see her from upstairs or even know that she was wanting something. Stella could have her needs attended to much more quickly if she could tell us exactly what she wanted.

After saying her first independent word, Stella reached the groove of saying all three of her words by herself. She said "outside" nearly every time she needed to go to the bathroom or wanted to go out to play. Jake and I were thrilled that this meant way fewer accidents in the house. She said "water" whenever she noticed her bowl was empty.

I was most impressed with how Stella started using "play" on her own. She could play with her toys whenever she wanted. We kept her basket of toys on the ground, easily accessible to her at all times. But Stella started saying "play" during times when Jake and I were focused on other things besides her, like when we were eating dinner, or in the middle of a deep talk, or on our computers. If we failed to drop whatever we were doing when she approached us with a toy dangling from her mouth, she would go say "play" and trot back over to us. Some nights, Stella said "play" between ten and twenty times. Words were becoming one other way she could communicate her point. They were another tool in her expressive repertoire.

One evening in early June, Stella sat in front of the stove watching me cook dinner. I could feel her eyes tracking my every movement. Every few seconds she inched closer to me, making sure I was aware of her presence. Suddenly, a roaring sound came from the front yard. Stella dashed into the living room, leaped onto the couch, and peered out the window to see Jake mowing the front lawn. When he disappeared from her sight, Stella barked three times, then jumped down. She wedged her head between the shade and the window on our front door. As soon as Jake came back into sight, her tail wagged and she barked again. Stella ran back through the kitchen, said "outside," then bolted to the front door. She stood on the doormat and looked back to me, checking to see if I heard her request and followed.

Of course, I was going to drop everything in the kitchen to take her out to the front yard, even though I knew she hated the lawn mower. This was the first time Stella had generalized "outside" to a location other than the backyard. I could not have been happier. If Stella thought that all three of her buttons were the same and she just pressed whichever one was closest to her to grab my attention, she would have chosen to hit the "play" button literally two feet away from the front door. Or she could have picked the "water" button that was on the way to the kitchen. But she did not do either of those. She ran to the opposite end of the house, completely out of her way, to tell me "outside."

Stella charged through the front door. She alternated between barking at the lawn mower and retreating to my legs to protect me. When Jake shut off the horrifying sound-making machine, she wiggled her entire body in celebration. All Stella wanted was to go outside to see Jake (and possibly save his life from the lawn mower). She had no idea that she achieved a huge communication milestone, and I had glimpsed her potential.

When I write speech therapy goals for emerging communicators, I always include at least one goal for the child to use language for a variety of communication functions. Requesting is often the most heavily targeted communication function with young children, but we need to remember *why* we all actually talk in the first place. We communicate to request items and actions, reject or protest, comment on what is happening, share ideas and feelings, ask questions, answer questions, label objects, direct the actions of others, joke, tell stories, and so much more. When I had the initial idea to teach Stella how to say a few different words, I envisioned her using them to make different requests, to let us know what she wanted or needed. It did not even occur to me to think about supporting Stella's different communication functions until later in June, the day before we were leaving for our family vacation.

I watered my houseplants with Stella standing by my side. Ever since she was eight weeks old, Stella was fascinated by this weekly activity. She followed me from room to room, plant to plant, her eye gaze shifting back and forth from the spout of the watering can to the soil. This time, she left after I finished watering the first plant. About five seconds later, I heard, "water." I chuckled to myself. I bet seeing the water reminded her that she was thirsty. I started walking down the hall to see Stella walking back toward me.

"Do you need water, girl?"

As I saw moments later, her water dish was full, and she did not take a drink. Stella returned to the sunroom with me, continuing to watch me give the plants a good soak. She was talking about what was happening, not about something she wanted.

This was the first time Stella said a word without requesting an object or an action from one of us. She walked out of the sunroom, down the hall, and into the dining room all simply to tell me what she observed in the world around her. It certainly was not easy or convenient for her to share her thoughts, but she did so anyway. I texted Grace immediately relaying the story to her. What other communication functions could Stella use? What else was she noticing in her world that she did not have words for?

The next day, Jake, Stella, and I road-tripped to a lake house in Wisconsin for my family's annual summer vacation. My parents rented the same house each year for the now eight adults, two dogs, and one baby in our family. The front door faced the narrow dirt road surrounded by miles of deep woods on both sides of it. The back door opened to a large, grassy yard with stone steps leading to the sandy shore of Lake Michigan. The space was the most serene bubble away from reality. We spent most of our days relaxing, reading, chatting, and enjoying nature. At night, we would play board games then roast marshmallows in a bonfire pit under the stars. There was little to no cell service, and no internet or cable. This week was always a chance for all of us to take a time-out from our busy lives, reconnect with each other, and let our minds recharge and wander.

This year, I had even more to look forward to. Jake would be on this trip with me for the first time, and we would introduce Stella to her first beach. I could not wait to see how she would react to the sand, water, and waves. Plus, I was eager to see how Stella would communicate in a new place. I wondered if being in an unfamiliar house with different people would impact how she used her words. In speech therapy, it can take a while for a child to transfer what he learns in the treatment room to other environments. I set Stella's buttons up right away when we arrived and modeled each of them a couple of times for her. The "outside" button went next to the back door. "Water" went by her dishes in the hall across from the back door, and "play" sat in the living room near her toys. But, after placing them down initially, I did not pay too much attention to using her buttons. I was preoccupied with catching up with everyone, playing with my five-month-old niece, and enjoying the beach.

Jake and I took Stella out for a walk along the shore. The

water level was higher than I had ever seen it. The lake practically swallowed the shoreline, leaving just enough room for two people to walk on the sand.

"Let's let Stella off her leash," Jake said.

I was extremely hesitant to try this. I had no idea how Stella would react. My family used to let Wrigley off-leash here, but she was much older. I could think of a thousand reasons to say no. There were woods in the distance, we had not practiced this even in a normal park, we didn't have any treats with us to lure her back if she sprinted away, there was no cell service to call for help if we lost her, and she was still so young. But somehow, Jake convinced me that he could run faster than Stella if she sprinted away, and that the two of us would keep her safe. Jake unhooked her leash. I took a deep breath in and crossed my fingers. Stella was officially free for the first time out of our backyard.

Much to my surprise, Stella continued trotting along right in front of us. "Good girl, Stella. Good girl." Stella looked back to Jake and me. She smiled and panted, then bolted ahead. She was at least thirty feet in front of us with no signs of slowing down.

"Stella, wait," I shouted. I shot Jake a look, cueing him to go ahead and start running to catch her.

But Jake did not need to sprint ahead. Stella stopped dead in her tracks. She turned her head back to us and continued smiling.

"Good waiting, Stella. Yes, good girl," I shouted. Stella stood still until we caught up to her. She wagged her tail and jumped up to greet us, as if we had been gone for hours.

I was shocked. Stella was a high-energy puppy, on the beach and off-leash for the first time in her life. She cared enough about us to stop exploring her wide-open surroundings, turn around, and wait for us to catch up to her. Stella continued trotting by our side for another few seconds, then ran ahead again. As soon as I called out for her to wait, she did not take a single step farther until we reached her. This became our pattern for the rest of the walk back to the house.

I wondered if she was listening to us so well because we listen to her so much. It's the same way with kids. The more I pay attention to and really listen to their communication, the more they'll notice and listen to mine. Jake, Stella, and I were building a strong relationship together, based on presuming competence, listening to one another, and trying new things. Every time we acknowledged one of Stella's gestures, vocalizations, barks, or words, we were modeling how to react to another's communication. Now, Stella stopped and listened when we talked too.

If it were not for Jake's push, I never would have given Stella the chance to show us how well she could do. We had truly only practiced the word *wait* a few times at home. We would tell Stella to sit, then say "wait" repeatedly while we slowly backed up. We had only made it across the living room before Stella came running over to us looking for a treat I clenched in my hand. This situation on the beach was different. It actually mattered to Stella and made sense why she should wait. We were in a new environment and she was actually pretty far away from us. This was even more proof to me that both dogs and children learn and perform best when it is meaningful to them. There was not a good reason to have Stella sit in the middle of the living room until we disappeared out of sight and called for her again. She knew we were all still in our home, and she was just waiting for a treat.

A couple of days into our vacation, Stella became comfortable with her new surroundings and all the people who could give her lots of playtime and belly rubs. She was truly a ham, soaking up all the attention. My family all gathered to eat lunch around the dining room table. Stella wandered around the first floor, continuously coming in and out of the kitchen. In the middle of our meal we heard "outside" come from the back-door area. Everyone quieted down and looked around the table, taking a quick mental head count. We were all there together. That meant the only one who could have said "outside" was Stella. Sitting closest to the door, my dad laughed when he leaned back in his chair to see little Stella standing patiently by the door, waiting to go outside.

"I'll take her," he said, laughing. "Yes, that's a good girl, let's go outside." His voice trailed as he walked out the door.

Throughout the rest of the week, Stella continued to say "outside," "water," and "play," without missing a beat. Until this trip, I had no idea if Stella would use her buttons in other locations, or if her learning was specific only to our home. The fact that Stella used her words here in the same ways as she would at home encouraged me. It was one more indicator that Stella was demonstrating true, independent communication.

"If you could live anywhere in the country, where would you go?" I asked my sisters, my mom, and Jake. We all sat on a deck overlooking the lake.

Jake and I had begun exploring thoughts about moving. We loved Omaha but felt that it was time to go. We were ready for new jobs, neither one of our families were in Omaha, and several of our close friends had moved away. Most importantly, we had an unfulfilled sense of adventure that kept creeping up to the forefront of our minds whenever we allowed it to.

Something hit me after becoming a fully licensed speech therapist. I had arrived to the stage of my career where there were no more built-in stopping points like there were in school. I had to make my own stopping points if I wanted anything to change. For college, graduate school, and my clinical fellowship, I picked where to go based on the school or job. This time, I wanted to pick where I wanted to live, then find the perfect job. If not now, then when?

"Wherever we go next, I want there to be a body of water," Jake said.

"Maybe we'll move to Milwaukee. The lakefront is beautiful there," I said. Stella sat in the sand, looking out to the waves, her nose sniffing the air around her. "Do you want to be a beach dog, Stella? I bet you'd like that."

Moving somewhere new wasn't the only big thing on my mind. Stella's communication skills were already surpassing where I initially thought they could go. At four months old, Stella was using three words independently, in multiple environments, with several different people. This was incredibly impressive to me. I could not believe how fast she learned. During the process, it seemed like it took so long, but now, looking back, we only had Stella for two months and she was already saying three different words. We could hear her desires from anywhere in the house, and she rarely went to the bathroom inside now. She did not have to wait for me to guess if she wanted to go out. But to me, *how* Stella learned to use these words was so much more impressive than what she was saying at this point.

When I set out to teach Stella words, I asked the question, *What happens when I implement speech therapy language interventions with my puppy?* I did not ask, *How can I get her to push these buttons that say words?* I am sure there are lots of ways someone could train a dog to push three buttons when they want the dog to. But I taught Stella in all the same ways that I teach children. I didn't train her to push a button on my command. I didn't reward her with treats for saying a word. I didn't use my hand to grab her paw and make her say a word. I didn't train the behavior of pushing the button first before I attached meaning to it. Those approaches would not have been in alignment with best practices for teaching children words.

Stella's ability to learn quickly and with such ease shed light on a power of speech therapy that I imagined reached far beyond where others knew it could go. Now the parallels I saw between Stella's and toddlers' prelinguistic skills were continuing into the linguistic phase. Focusing on *how* we arrived here meant that there was so much more potential left to explore. If the communication similarities between puppies and toddlers continued to this point, what other words and concepts could Stella learn? I couldn't stop here. We were just getting started.

I came back to Omaha knowing two things: (1) I was ready to leave and start a new life somewhere else; and (2) Stella needed more words.

TAKEAWAYS FOR TEACHING *YOUR* DOG

- **Provide a long pause.** When you see your dog noticing your modeling or noticing the buttons, turn your routine interactions into language-facilitating opportunities. The greatest cue we can provide is a long, silent pause to give the AAC user a chance to process what is happening and try exploring her words. When you see your dog communicate through a gesture or vocalization, stay quiet for at least ten to fifteen seconds. At the end of fifteen seconds, if your dog looks like she might be walking toward her buttons or is looking at them, continue staying quiet. If you have not seen an indication that she might try saying a word, add a naturalistic cue.
- **Your dog may need cues for a little while before using words independently.** Keep providing a long pause, pointing at the button, asking a general question such as "What do you want?" or standing near the button to support your dog's emerging vocabulary. Even after you've heard your dog's first words, your dog will likely need support before using words independently and regularly.
- **Model words in different contexts to support generalization.** Your dog will learn to use words in different ways if she sees and hears you using words in multiple ways.
- **Remember that your dog is intrinsically motivated to communicate.** Resist the desire to offer a treat for saying a word (unless the word is *treat*). This will keep your dog from learning the actual meaning of the word. Stick to providing the appropriate response to your dog's word.
- **Think about other communication functions besides requesting.** Your dog might be trying to label an object or activity in her environment or talk about what is happening.

CHAPTER 8

Project Disconnect

I returned to Omaha feeling refreshed, reinspired, and ready to make some big life changes. I wanted to choose a new city to live in based on what Jake and I wanted out of life, not based on a job. Jake and I could both find work in whichever city we chose. I craved adventure, balance, and free time. I wanted to let my mind explore, wander, and create like I had on vacation instead of constantly consuming television or social media. I looked down at my phone screen to see pages of apps with red notifications shouting at me. I caught myself mindlessly scrolling whenever I had downtime. I automatically turned Netflix on after dinner nearly every night. This is not what I wanted to become. I wanted to be deliberate about how I spent my time. These habits had not bothered me before. I didn't even realize

I had them. Now, after unplugging for a week, I questioned if I even wanted to allow them back in my life at all.

"What about San Diego?" I said. I stared down at the list of criteria Jake and I made for wherever we would move next. We spent this stormy Sunday inside, dreaming about our future together. *How did we want to live? Where did we want to go? What would our ideal place to live look like?* were the most important questions on our minds. "There are tons of outdoor activities, lots of road-trip destinations around it, it's seriously amazing," I said.

San Diego is my definition of paradise. I visited with friends a year before and loved it. It was the first time during a trip that I decided I wanted to stay longer than I had originally planned. San Diego had everything I could want—a fun city, mountains, and beaches, and the weather pretty much stayed in the 60s and 70s all year long. It was nearly always pleasant outside. I imagined taking Stella for walks on the beach after work, going for hikes on the weekend, and taking road trips to explore other parts of California.

"I think it's definitely in the top three," Jake said. We agreed to see what job options existed and go from there.

Three days later I had a video interview with the owner of a speech therapy company in San Diego. This job would be very different from my current position. I would be an early intervention speech therapist, working only with one- and two-year-olds with speech and language delays. I would provide services in each child's home instead of in a clinic like I was in now. It stimulated my sense of curiosity and adventure within my profession, gave me the power to create my own schedule, and I could take more days off than I could now. I had a great feeling about it as soon as I saw the job listing.

"She offered me the job," I said after closing my laptop.

"You're kidding me, right?" Jake was shocked by how quickly I move when I am set on an idea.

"There's no pressure to decide now. She knows it wouldn't be for a couple of months. But if we did move to San Diego, I would totally work for her."

I hoped if we changed jobs and moved, I would have more time to dedicate to teaching Stella. Since coming back from vacation thoughts about her progress and potential kept bouncing around in my head. Before Chaser, the border collie who learned the names of more than a thousand toys, research showed that the average dog understood 165 words.[19] Chaser taught us that dogs may have much greater lexicons than we thought possible. In humans, our receptive language and expressive language skills are typically pretty close to equal. We all usually understand more words than we use, but the two abilities should be in the ballpark of each other. I wondered, *So if dogs could understand hundreds or even upward of a thousand words, does that mean they could potentially say that many words too?*

Stella was learning so quickly, but I could barely contribute any time or mental energy to support her now. By the time I came home from work every night, it was already 8:00. I was exhausted, hungry, and eager to relax. She was receiving the fried version of myself teaching her. What could she achieve if I could really spend a lot of time working with her? What could she say if I could practice new words with her consistently every day? I felt like this was the prime time for her learning, yet I could not fully take advantage of it in my current circumstances. I hoped I wasn't losing my window of opportunity for her to learn new concepts. I resolved to keep teaching and keep learning as best I could. I ordered two more boxes of buttons and eagerly anticipated introducing more words.

The next morning, I sat in the sunroom enjoying a quiet few minutes of reading before I headed upstairs to shower. I could hear Stella's collar jingling and toy squeaking in the living room. She must have been running around playing. Stella's ability to entertain herself independently was increasing as she became a bit older.

The noises stopped. The quiet was always much more concerning than hearing what Stella was doing. She could be getting into anything or going to the bathroom in the house. I closed my book, about to go check on her. A few seconds later, Stella charged into the sunroom. She stopped in the doorway and barked at me.

"What do you want, girl?"

Stella's nostrils flared. She sighed and barked at me again, then ran back into the living room. I followed to see where she was taking me. Stella stood in the middle of the room looking all around, then back to me.

"Play? Stella want play?" I said "play" with Stella's button and tossed a toy across the room for her.

Stella looked unamused. She stayed standing in the center of the room. "No play? Okay, girl."

I returned to my book in the sunroom. I had only made it through about two sentences when I heard a scratching noise coming from the living room. I ran out to see Stella pawing the floor underneath the couch.

"What's under there, Stella? You need help?" I lay on the

ground and reached as far as I could under the couch. Stella shoved her nose right next to my arm. Neither of us found anything. I walked around to the back of the couch and crouched down. I shined my phone's flashlight into the narrow space between the couch and the wall. The corner of Stella's plush pizza toy was sticking out from under the back of the couch. "How did you even get this back here, Stella?" As soon as I pulled it out, Stella grabbed it from my hand and jumped on the couch to happily resume chewing it.

Stella needed my help when her toy slid behind the couch. Since I did not see what Stella was playing with or what happened, I had no idea what I was looking for. It would be so much easier for all three of us if Stella could say "help" when something like this happened. When the two additional boxes of Recordable Answer Buzzers that I ordered arrived, I thought about what Stella communicates to us with her gestures and vocalizations, which words we say to her frequently during routine activities, words that would allow her to communicate for functions other than requesting, and words that could be used across multiple contexts.

Word choice is a big deal in the world of AAC. The words we program or keep available for children are the only words they will be able to say. There are two categories of words: *core* words and *fringe* words. Core words are the most frequently used words in communication. Studies in which researchers analyze language samples from different populations and contexts have shown that there are approximately three hundred to four hundred words that make up about 80 percent of everything we say.[20] Core words are typically verbs, adjectives, pronouns, adverbs, and prepositions. The most effective AAC systems are set up to teach core vocabulary first "because it allows communicators to express a wide variety of concepts with a very small number of words. Since core words make up the majority of spoken language, focusing on core vocabulary allows many opportunities throughout the day to hear the same words being used in a natural environment."[21] Using core words sets the learner up for their best communication success.

I chose to add six buttons with core words and a phrase for Stella: *come*, *no*, *love you*, *help*, *bye*, *eat*. We told Stella to "come" all the time. I wanted her to be able to tell us "come" as well if we were in a different room. We said "no" to Stella if she was chewing on a belonging of ours, or doing something we did not want her to do. Stella deserved the opportunity to tell us when we were doing something she does not like too. Communication is a two-way street. Stella's comfort and happiness mattered to me. Jake and I always said "love you" when we cuddled with Stella, rubbed her belly, or kissed her head. I wanted to give Stella the chance to express her affection to us, or to others, as well. "Help" could be beneficial for Stella when she needed us to help her retrieve a stuck toy or even if something more serious happened. Every time we left for work, we told Stella "bye" multiple times. Our leaving on weekdays was a routine for her now. I wanted to give her a way to talk about that part of the day. And Stella already understood the word *eat*. Before feeding her breakfast and dinner, we always said, "eat." Stella licked her lips and ran to her bowl. Hopefully Stella could tell us when she wanted to eat breakfast and dinner instead of us completely creating the routines for her. Maybe she was hungrier earlier in the day, or maybe she was not quite ready to eat when we put food in her dish. We could not know these answers until Stella had a way to tell us herself.

The remaining 20 percent of vocabulary we use, the fringe

words, are more specific. They are typically nouns that can only be used to mean one thing. Everyone needs both core and fringe words to communicate effectively. I always recommend that a solid vocabulary consisting of mostly core words and a few fringe words be established first so the AAC user can communicate throughout all of his day. More fringe words can be added later to help learners be more specific in their communication. *Play* is an example of a core word. It can be used to talk about all toys and games, whereas *ball* and *toy* are fringe words. *Eat* is a core word. It can be used to talk about all foods and mealtimes, whereas *breakfast*, *dinner*, and *peanut butter* are specific fringe words. Stella hears "eat" many more times in any given day than she hears "dinner."

I included one important fringe word to Stella's available vocabulary: *walk*. We took Stella on walks almost every night after dinner. It was a highly motivating activity for Stella and a significant part of her day. Like most dogs, Stella became so excited every time we asked if she wanted to go for a walk. She would bark, jump around in circles, and stalk me until I followed through by putting my shoes on, grabbing her leash, or finding my headphones. She used every type of communication possible to tell me that she wanted to go for a walk. I wanted Stella to be able to request walks instead of having to wait for us to say one of her favorite words.

For now, I did not want to become caught up in the number of words Stella could say. I knew that with a good selection of vocabulary, she could likely communicate about most parts of her day by using core words. There was no way we were going to place hundreds or a thousand buttons on the ground to experiment with every word she might know. Kids have access to thousands of words on their devices and use the same few hundred most commonly. The best AAC systems are arranged so that core words are on the front page, and fringe words are on the second and third pages. This way, kids can have easiest access to the words they need to say most often.

I wondered if Stella would try pressing her new buttons right away, or if it would take a while for her to notice them. Was the monthlong modeling time frame normal, or was it longer because they were her first words? Would Stella use her old words less and be affected by the novelty of her new vocabulary? I had so many questions that I was eager to start exploring.

I programmed six buttons for her new words and found spots for each button. I set "eat" down next to her food dish, and "bye" next to the front door. I lined up "come," "no," "love you," and "help" on the floor in the living room next to our entertainment center.

"Stella, come," I said verbally and pushed her new "come" button.

Stella trotted over to me.

"Look, girl." I pointed to each of the four buttons. "Now you can say, 'come,' 'no,' 'love you,' and 'help,'" I said while pushing one button at a time. Stella stood next to me. She switched between licking my face and looking down at the buzzers.

After figuring out the best placement for her new buttons, I would always keep them in the same spot. Stella would learn fastest if she didn't have to search for the word she needs every time she wants to say something.

The most effective AAC systems are set up by the principles of motor learning and motor memory.[22] Words and phrases should always stay in the same location on the device, so the locations become automatic to the user. It's similar to how we learn to type on a QWERTY keyboard. If the keys were rearranged on every different keyboard, we would have to search for each and every letter. But since the keys have always stayed in the same location, our fingers automatically know where to go as we type. We do not have to think about where a letter is located at the same time as typing. Having to search for words or letters takes up mental energy that could be spent on communicating.

I saved the best word for last. It took me several tries of shouting "walk" into the blue buzzer to make the final /k/ sound audible. Stella came running over to me when she heard me saying one of her favorite words so many times in a row. When I finally recorded it just right, I placed the "walk" button under the hooks next to the front door where Stella's leash hung.

"Look, Stella, walk!" I pushed her button a couple of times in a row. Stella stood hovering over the buzzer. She cocked her head to the left and to the right. She wagged her tail and looked back up to me.

"Are you excited, girl?" I crouched down to Stella's level. She licked me repeatedly and turned around to face her new button again. Stella lifted her right paw, swatted down, and said, "walk," only one minute after I programmed it for her.

"Go for a walk? Okay, let's go."

Stella barked and smiled at me while she watched me put my shoes on and grab my headphones. "Walk," I said one more time verbally and with her button. "Let's go."

It was fascinating to see such a clear difference in her reactions between all her other vocabulary words and "walk." Stella did not push any of the other words right away, not even "eat." She stood near me while I modeled them, but they did not capture her attention like "walk" had. Seeing her so excited to push it made me wonder if she had ever wished for the ability to say "walk" before I programmed it for her.

The next day, Jake and I sat down at the dinner table, exhausted from our days at work. I brushed the piles of mail, papers, and clutter down to a chair to make room for our plates. Stella was lying down in the living room chomping on a chew toy. I was ready to relax with a nice, calm dinner and a glass of wine. We made it through about two minutes of our meal when we heard, "Walk walk walk walk." Stella poked her head around the corner, checking to see our reaction to her request.

"No walk now, Stella. We're eating." I laughed.

Stella maintained eye contact with me, then barked.

"I hear you, Stella. We'll go for a walk later!"

Stella disappeared from our view. A few seconds later we heard, "walk, walk," followed by a bark.

"It's her second day being able to say 'walk.' She's so excited! Should we take her now and finish eating when we get back?" I asked.

Jake chuckled. "Of course she had to choose now to say a new word a bunch of times. Let's go . . ."

After three days of having seven new words, Stella started saying "walk" all throughout the day without any cueing from us at all. I knew she liked walks, but I had no idea how often she wanted to take them. I noticed that Stella started saying "outside" less often after we introduced "walk." She still said "outside" when she needed to go to the bathroom, but this change in frequency made me wonder if some of the times Stella had been saying "outside," she was wanting to go for a walk. Until now, she could not have specified that with a word.

Stella had not used any of her other new words yet. The closest she had come to exploring them was when her toy bounced and landed in between the buttons by our entertainment center. She pawed at her toy, but accidentally pushed "love you" in the process. Stella cocked her head. She looked surprised to hear "love you."

"Aw, love you, Stella!" I kissed her forehead and scratched behind her ears. Stella licked my face. She pawed the button instead of her toy by accident, but it is important to always respond as if it were intentional. That would help her learn what each word means. If I had ignored Stella accidentally saying "love you," she would have learned, "I pushed this button that said, 'love you,' but nothing happened." That would likely decrease her chance of pushing it intentionally in the future. Instead she learned, "After I pushed this button that said, 'love you,' I received scratches and attention."

Accidental hits, mis-hits, and pure exploration are all fantastic opportunities for AAC users to learn their vocabulary.[23] The more Stella sees us respond appropriately after she says a word, the faster Stella will learn what that word means. At work, anytime a child said a word with his device, even if I knew he was looking to say something else or bumped it by mistake, I always responded to the word he said. Mis-hits are valuable. They provide great opportunities for the learner to hear what they said, observe how their communication partner responded, and decide if that is what they actually wanted or not. If it was not what he intended, the AAC user can try a different word. He would then be exposed to the responses of multiple words instead of only one.

Jake and I had officially decided to move to San Diego. I accepted the early intervention position, and Jake was deciding between a couple of different job offers. We handed in our thirty-day notices at work and started looking for apartments in California. Since we had so much to do to prepare for our move across the country, I figured now was as good of a time as any to prune out the habits that I felt were holding me back.

By the end of July, about a month after we returned from vacation, I still fantasized about unplugging from the overly connected world. I had slipped back into my routines of mindlessly scrolling through social media, unintentionally taking up all the idle moments of my day. Ever since high school I had logged on to at least one social media platform nearly every day. It was so ingrained in our culture, so habitual. I never asked myself if I wanted to, I just did. But I rarely felt great after spending time online. It felt like I was filling my brain with the thoughts, ideas, and photos of people I barely spoke to anymore. This could not be healthy. It certainly was not benefiting me in any way that I could see. I kept wondering what it would feel like to not be a part of that anymore. Would I feel like I was missing out? How would I learn about my friends' big life events? How would I share photos with friends and family members if I wanted to?

One night after work, Jake came downstairs to find me trying to lift the fifty-inch TV off the wooden entertainment center that he built.

"Whoa, what are you doing?" he asked.

"I unplugged the TV! And I deleted all social media apps from my phone, too. I'm done wasting time. I'm calling it 'Project Disconnect.' Let's try putting the TV in the basement. Can you help me carry it?"

"Don't you think that's a little extreme? We can just choose not to watch TV," he said. We both loved to watch shows at night.

Jake and I compromised. We placed the TV behind the console. It was out of sight for me, and still easily accessible if we changed our minds.

Deleting so many apps from my phone brought a couple of significant realizations. Now, all the icons were in a different order. I had to visually scan the screen to find what I was looking for. Even though there were many fewer apps and I only had one page of icons instead of three to look through, it took me longer to find each one. Research shows that motor plans are stronger than recognizing a visual cue for an icon.[24] Before now, I never had to think about where my frequently used apps were. My finger just went there automatically. Experiencing this struggle after changing my setup reminded me how important it was to keep words in the same location for AAC users.

Most importantly, after a couple of days without using social media or watching TV, I already felt a huge difference. I had more energy at the end of the day, I realized how much more time I had when I did not take up every spare moment, and I was so excited. Since my brain was not filled with other people's thoughts, I could really listen to my own.

TAKEAWAYS FOR TEACHING *YOUR* DOG

- **When deciding which words to teach, ask yourself these questions.** *What does my dog communicate with her gestures and vocalizations? Which words do I say to her frequently during routine activities? Which words would allow her to communicate for functions other than requesting? Which words could be used across multiple contexts?*

- **Establish a vocabulary with more core words than fringe words.** For the most communication potential, and more complex use later on, teach high-frequency words.

- **Keep your dog's buttons in the same spot.** We all learn how to talk through the principles of motor learning. Verbal speech users learn the motor plan to say each sound and word, sign language users develop a motor plan for each hand movement, and AAC users for each word location. Continuously moving words around will be confusing for your dog. After you find a spot that works well, stick with it.

- **Spend time modeling new vocabulary.** Whenever you add more words, make sure to model them in their appropriate contexts.

- **Respond to accidental hits, mis-hits, and your dog exploring the buttons.** Even if you think you are absolutely positive that your dog pushed a button accidentally or meant to push a different one, respond to the word she said. These are valuable teaching moments. Plus, your dog might surprise you with what she says and when. To help your dog become a deliberate communicator, always respond as though the message was intentional.

- **Give your dog more words than you think she knows.** This allows for language growth and exploration.

CHAPTER 9

Bye, Omaha

Most people assume dogs would say "eat" all day long if they could talk. But this has never been the case with Stella.

Even though I knew Stella had much more to say than requesting to eat, I was still surprised when *eat* was not one of the first of the newly programmed words that she started using. I thought *eat* would provide the easiest and strongest correlation between a word and an outcome for her to make. That is one of the main reasons why I purposefully did not select *eat* as one of Stella's first few words. I thought that the food connection would be too powerful, that it would overtake her ability to learn that she could talk for reasons other than wanting food or treats. Or I thought that Stella would think all her buttons were simply a mechanism to gain more food from us. But I could not have been more wrong. I underestimated Stella's motivations and desire to connect with us about many other activities and parts of her environment, not just food.

When Stella did not say "eat" immediately, I started questioning why I thought that would even be true in the first place. Stella certainly enjoyed eating, but she never sat by her bowl, pawing at her dish all day long. If nonverbally she was not communicating this desire constantly, why would I expect her to suddenly switch to only talking about food now that she could use a word to say "eat"? Stella was interested in and driven by much more than food. She loved being outside, going for walks, playing with us, and receiving positive praise. She watched us closely throughout many events of the day, not only mealtimes. And, like all dogs, Stella communicated nonverbally and with vocalizations about much more than wanting to eat.

Stella picked up on "bye" second fastest after "walk." Jake and I said "bye" to Stella every time either one of us left, so she heard that word quite a bit. I wanted to give Stella the chance to tell us "bye" when we were leaving, to acknowledge the familiar event that was about to happen. Kids learn words best when they are incorporated into their typical activities. Routines are predictable, frequently occurring, and functional.[25] Since Stella also thrived with incorporating structure into her daily life, I figured the same might be true for her as well. I also wanted to introduce "bye" because it represented a new category of vocabulary for her: social words. *Hi*, *bye*, *uh-oh*, *please*, *thank you* are all examples of toddlers' early developing social words. Along with nouns and verbs, social words make up a significant part of toddlers' utterances.

Late one Saturday night, we were in the middle of a stereotypical midwestern good-bye with a few friends who had been hanging out at our house. The midwestern good-bye involves at least a twenty- or thirty-minute-long process of announcing that you are leaving, making the rounds one more time, starting whole new conversations, then talking for another ten minutes by the front door. We all congregated there, had already hugged good-bye, yet continued our conversations. Stella had stayed up way past her typical bedtime playing with us and enjoying all the attention. She hopped off the couch and walked right through the middle of our circle. She looked up at all of us, then swatted her paw to say, "bye." She looked back up at our friends.

Everyone looked at us in shock and looked back down to Stella. "You want us to leave, Stella? Bye, girl!" one friend said. "I can't believe your dog told us 'bye.' This is wild, you guys," another friend said, laughing.

They left and Stella sat looking through the glass door as they walked to their cars. Then she hopped back on the couch, curled up in a ball, and fell asleep. I am not sure if Stella wanted them to leave so she could go to bed, or if she was commenting on what she knew was about to happen. But either way, it was so exciting to see Stella start to use another one of her words and share thoughts about what was going on.

Stella loved standing on the back of the couch and looking out the living room picture windows. Perched on her post, she could see all that was happening in the world in front of her house. She whined when she saw squirrels running up the tree, barked at strangers walking along the sidewalk, and jumped around in circles if she saw one of our cars pull up to the driveway. While I was in the kitchen, I heard the sound of the blinds clanking against the window. I hurried into the living room to see Stella pawing repeatedly at the closed shades. She was gesturing that she wanted the blinds pulled up so she could look out the window.

I joined Stella on the couch and tapped the blinds with my hand. I said "help," then walked over a couple of feet to also say "help" with Stella's button. When I pulled the shades up, I said "help" a few more times and watched Stella's gaze following the blinds sliding up the window. By imitating her gestures, I was showing that I understood what she wanted. Then I added a word to go along with her gestures.

In speech therapy sessions with AAC users, I recognize and accept all forms of their communication, not only to words spoken with the device. The more communication we recognize and respond to, the more the learner communicates overall. If the child points to something she wants, I also point to it, then model the word she could say. I am showing her that I see and understand what she is communicating, then say "want" or "get" or "help" to teach her a word she could use in that situation.

We do this naturally with babies all the time. When a baby waves to you, you likely instinctively wave back and say, "Bye" or "Hi." Or when a baby claps, you probably clap your own hands and say, "Yay!" Without even realizing it, you recognized the baby's gesture, associated a meaning to it, reinforced the gesture by doing it yourself, and modeled the word that goes along with that gesture. This is the heart of teaching language: making a conscious effort to recognize all forms of communication, responding to them, and modeling the next level.

Since *help* is a versatile core word, I modeled it throughout several different contexts—pulling the blinds up, finding a toy for Stella, putting her harness on, moving her dishes to find the one morsel of food that fell behind her bowl. Modeling words in varying ways helps the learner understand that each word can be used to talk about more than one or two different situations.

Stella started using her sixth word, *help*, multiple times a day when she would drop her toy behind the couch or when it rolled under the TV stand. Initially, she said "help" on its own, without pairing gestures with it. Even though we knew she needed help, if Jake or I had not been watching, we would not know where her toy disappeared. Stella would watch us look around, under the couch or under the entertainment center. Then, her tail started wagging and she came to stand right next to us when we were close to it. Soon, Stella started independently combining "help" with a gesture, normally standing where she wanted us to look. She learned that we needed more information to most effectively find her toy.

As Jake and I started seriously preparing for our fifteen-hundred-plus-mile move from Nebraska to California, it was clear right away that we would need to significantly downsize our belongings before we could leave. There was absolutely no way we could take all the furniture and belongings from a four-bedroom, two-story house to a small one-bedroom apartment in California. We considered all the possible moving strategies: driving a U-Haul with one of our cars attached to the back all the way through the mountains and national parks we hoped to visit on the way, shipping boxes of our possessions, or loading up a giant twenty-foot-long POD that someone else would drive out to San Diego. After days of weighing the options, neither Jake nor I felt like we landed on a solution that we loved. All the options were expensive and required a decent amount of planning. We had no idea what would fit in the small apartment waiting for us.

"What if you guys just packed up and drove out in your two cars and bought new furniture when you get there?" Grace asked.

This suggestion came at the perfect time. Jake and I had recently watched a documentary, *Minimalism*, which discussed the benefits of living with fewer material objects. Minimalism is the intentional promotion of the things that we most value and removing the things, mainly material objects, that distract from that.[26] The founders of the minimalism movement preached how letting go of nonessential items could make space for more creativity, inner peace, and purpose.

Learning about this concept opened our eyes to the amount of possessions requiring our continual attention and maintenance. Before now, it had really never occurred to me to purge all the nonessential possessions I carried with me through different stages of life and to stop automatically accumulating more. This could be an incredible opportunity for us to start fresh and let go of whatever did not serve us anymore. Jake and I could pick out furniture and décor that specifically fit in our new space.

Initially, Jake resisted. "Why would we get rid of things that we will just have to buy again?" he asked. But he hopped on board after he calculated how much everything we considered bringing was even worth. It would cost more to move it. "Let's go for it. We can sell our stuff here and use the money to furnish our new apartment," he said.

Now the idea spoke both our languages. It was a new personal growth challenge for me, and a practical, financially sound decision for Jake. Over the course of the next month, all our spare time was spent confronting the belongings we had accumulated and carried with us unintentionally for years. I fed my mind with podcasts, blog posts, and books about living with less and the process of letting go. As everyone recommended, we started with the easy things to part with—college textbooks, old DVD cases, clothes we never wore, old roommates' kitchen utensils. Clutter I did not even know we had crept out of the darkness, into the light of day for us to deal with. Then, we reassessed items of any sentimental value—knickknacks we collected throughout our travels, favorite T-shirts from high school, several-year-old gifts from friends or family that we no longer had a purpose for. Every time I resisted letting go, I reminded myself that if I continued on this way, I would never have room for anything new to come into my life or to appreciate my favorite belongings.

We took eight carloads of bins and bags filled with random household objects, furniture, and clothes to Goodwill. We sold our TV, four couches, chairs, tables, appliances, and decorations on Facebook Marketplace and Craigslist. Every evening after work, three or four strangers would show up to our front door to look at an item we were selling. Slowly as we stripped away everything, I started feeling relief. I could see more blank wall space, fit all my clothes in my closet, take a water bottle off the shelf without knocking over a mountain of Tupperware in the process, and could actually appreciate the beautiful hardwood floors and structure of the house. With every trip we made to Goodwill, our minds and our house felt lighter. I never had to worry about finding storage space for these random items any longer. I was freeing myself. My emotions swung from shock at how much I owned to the relief of letting it all go.

Stella did not adjust well to the purging. Her environment was being turned upside down right in front of her. With each piece of furniture that we walked out the door, Stella stood in the middle of the room, stunned. When Jake and a few friends carried out the couch, I stayed inside with Stella to make sure she didn't slip out. Stella walked into the bare sunroom and whined right where the couch used to be. Stella left the room, turned the corner, and said, "no." She looked up to me with the saddest eyes before she curled up in a ball on top of an empty grocery bag lying on the ground. This was the first time Stella had ever said "no." She was either commenting on how there was "no" couch there anymore, upset that the couch she grew up snuggling on was suddenly gone, or protesting what we did. Maybe Stella had more awareness and attachment to the items in our home than I had realized.

Jake and I were constantly buzzing around the house, packing up items we were taking with us, and cleaning out emptied shelves and cabinets. We barely had enough time to stop and rest.

One afternoon, I was upstairs listening to music and emptying out the linen closet. Stella was with me for a bit, but she had left a few minutes ago to go downstairs. I turned my music down to keep an ear out for what she might be doing.

"Come, come," I heard.

I went downstairs to see Stella standing in the middle of the living room. As soon as she saw me, she dove on her toy and ran away. She called me down to come play with her.

Later on, Jake and I were in the basement sorting through boxes.

"Come," we heard again. A few seconds later, Stella's head appeared at the top of the staircase.

"You want us to come up?" I said.

Stella wagged her tail while she watched us walk up the stairs. She immediately rolled over, gesturing for a belly rub.

"Is this why you wanted us to come up here, Stella?" Jake laughed.

I liked that Stella had a way to tell us she wanted us nearby, especially now that Jake and I were not hanging out in the same spots for long periods of time. Most of the time, Stella would follow us around, play close to us, or watch whatever we were doing. But sometimes, she wanted us to come to her. I always modeled "come" naturally whenever I called Stella over to us. By saying "come," and using her button for it, Stella learned how and when she could say it to call us over, too. I found it impressive and interesting that when we were in sight of Stella or were already paying attention to her, she would say "play" if she wanted us to engage in a game. But when we were not paying attention, or were out of sight, she said "come" to call us over to her. Stella never used "come" when she already had our attention, or when we were already where she wanted us to be. Observing the contexts in which Stella used each word gave me important information about her understanding of it. If Stella would have said "come" in all the same situations as "play" or any other word for that matter, it would be difficult to know if she understood the different meanings of each individual word.

Paying attention to how Stella used each word is a much more effective way to assess her language capabilities than quizzing her on the locations of each button. For example, testing Stella by asking "Where is come?" or "Where is play?" and waiting to see if she could find the right button does not actually tell us anything about her ability to communicate with it. It is an entirely different skill. It would be the equivalent of someone asking you "Tell me exactly where the letter 'g' is on the keyboard." Knowing that *g* is in the middle of the second row is not the same as knowing which words are spelled with the letter *g* or how to use it appropriately. So if I asked Stella "Where is come?" and she pushed her "come" button, that would not tell me Stella knew the word *come* or knew how to use it. It would tell me that she memorized its location, and possibly was trained to answer that specific question. A much better way of assessing Stella's language acquisition is tracking her word use along with the environment, context, gestures she used, and typical routines. It is always better to assess AAC users' language skills by analyzing their communication patterns.

After Stella had started saying five of her new words, *walk*, *bye*, *help*, *no*, and *come*, I deliberately gave her more pause time before I fed her each meal. It can be easy to go through motions on autopilot and forget to turn activities into language-facilitating experiences. We need to give emerging communicators the chance to speak up for themselves instead of always speaking for them. When Stella walked past her food dish and licked her lips after she woke up one morning, I stayed silent. She walked across the dining room to sniff the drawer her food was in. Stella walked back to the side of her food dish, said "eat," and looked back to me.

"Stella eat! Yes, Stella eat now."

Stella licked her lips and wagged her tail while she watched me carry her cup of food to her bowl. That night, Stella said "eat" again ten minutes before her normal dinnertime. Stella had now used every new word, except for *love you*, independently and consistently in the appropriate contexts.

Having more words available did not stop Stella from using her old vocabulary. She started communicating about even more activities throughout her day. When I introduced this second set of words to Stella, a lot of her learning came from exploring the buttons and watching our reactions. With all the chaos of emptying our house and preparing to move, I did not model language nearly as much as I had over the first few weeks with Stella's first words. But she had a strong foundation and was able to keep learning on her own. Stella knew that the buttons were for communication. After she heard the words and saw what happened when she said each of them a few times, she started using them more confidently.

Our move grew closer and our house emptier and emptier. Now our voices echoed in most of the rooms, and Stella's buttons, toys, and bed were our only décor. I kept Stella's favorite ball, plush toy, and rope toys out for her to play with over the next couple of days and packed the rest in a suitcase. Stella watched me zip up the suitcase, said "help," and curled up in a ball on my lap. Ever since she witnessed us packing up her belongings, she did not leave my side. I hoped she was not afraid we were leaving her. I hated seeing Stella so upset. I wondered what she thought was happening. Even I was feeling nervous about everything, and I knew exactly what was happening. I could not imagine what Stella was feeling while seeing her routines crumbling and all our belongings disappearing without knowing why. I wished I could communicate to Stella that we were going to be in a new home with new furniture and all her toys soon.

Over the course of a month and a half, Jake and I went from

having every cabinet, closet, and room overflowing with objects to small stacks of our suitcases, houseplants, and a few boxes by the front door. It was exhausting.

I went to bed with my stomach in knots. The next morning, we would hop in the car and leave behind entire lives we built here. Omaha was familiar and kind to me. I knew my way around, had favorite restaurants, bars, and plant shops. I had professional connections, a network of people I could reach out to. Jake and I met here. I had found a close-knit group of friends and could easily drive home to see my family in Illinois. I wondered how our lives would change, if we were making a huge mistake. But there was no going back now. We had new jobs to start, a new apartment to move into, a new life waiting for us to come and claim it.

TAKEAWAYS FOR TEACHING *YOUR* DOG

- **Add a word to your dog's gestures.** Model the next level of communication by saying the word that corresponds to your dog's gesturing. Your dog is learning another way she can communicate that concept.
- **Look for patterns in your dog's vocabulary use.** Learn more about your dog by paying attention to the words she uses most frequently and the specific situations in which she most often uses words. Just like people, every dog is different and will have different communication trends.

CHAPTER 10

The Road to California

Stella sat on her bed in the front seat of my car, a smile stretched across her face. She switched back and forth between sitting up, looking out the window to the never-ending cornfields over the rolling hills of Nebraska, and lying down on the patch of sun that hit her blanket. My copilot and I drove into the future with a bright blue sky above us. I have always loved being on the road. I feel my freest when I'm behind the wheel, listening to music, with nowhere else I need to be, or nothing else I could possibly be doing. It's a meditative experience.

Jake packed our cars like a game of Tetris, each object perfectly locked in its place, except for the grocery bag of Stella's buttons on the floor of the front seat. It was the only place we could keep them that we would be able to easily take them out. Whenever I stopped, turned, or hit a bump, I heard a rattling of plastic and "outside water walk love you help" at the same time. Stella looked down, then looked over to me.

"Yep, I heard it, too, Stella." This was going to be a long thirty hours.

We arrived in Breckenridge, Colorado, after our first nine-hour stretch of the drive. Stella was a perfect passenger the whole way, but as soon as we left the car, she exploded with energy. When Jake and I flopped onto the king-size bed, Stella started running laps around the hotel room. She jumped on the bed, barked at us, hopped off, and ran around in circles by the door. So far, Stella's first night ever in a hotel room was complete chaos.

"Oh, jeez, girl, you're so wound up," I said. Jake placed Stella's buttons around the hotel room. I tossed a toy for Stella and she dove across the floor, knocking the buzzers Jake just set up in all directions. I heard the beeps of a few of them resetting after they banged into the wall.

"Well, this is off to a great start," I said. "What if she's this crazy every night? We're never going to get any sleep." I set the buttons back up.

"Outside," I woke up hearing the next morning at 5:30 A.M. I could only see the outline of Stella standing by the door in the pitch-black hotel room. I stayed in bed while Jake took her out for a short walk to go to the bathroom.

We would need to start getting used to this. At our apartment we wouldn't be able to let her out in the backyard like we could in Omaha. Normally, she asked to go "outside" at least ten or fifteen times a day. It was now occurring to me that if she kept up that frequency in San Diego, we would be spending all our spare time taking her out. When Jake and Stella returned, she jumped on the bed and licked my face until I sat up. She hopped off the bed, took a drink of water, then said "eat" like she would at home.

"Stella eat! Okay, let's get your breakfast." Now I was energized for the day. It excited me to see Stella communicating with her buttons so similarly to how she would in Omaha despite the new location and stress we all had been experiencing for the past few days. Her words must have been becoming pretty automatic for her to say. If she really had to think about what to say or how to say it, she probably would not have used her AAC during this hectic time. It would have taken up too much mental energy for her. It's harder for everyone to do something challenging when they are stressed out. Dogs are no exception. Having access to her words seemed to settle Stella down a bit. She still had an abundance of energy, but she was not nearly as frantic as she was when we arrived in the hotel room before placing her buttons out for her.

On the second day of our drive, we traded the cornfields of Nebraska for mountains in Colorado and Utah covered with trees starting to show off their autumn yellow, red, and orange leaves. That night, we stayed in a small town in Utah. The single-story, U-shaped motel had about twenty rooms, each one accessible from the outside. When Jake and I were brushing our teeth, we walked out of the bathroom to see Stella peeing inside next to the front door.

"Shoot, I forgot to bring her buttons in from the car," I said. "That was my fault." Stella had been hanging out by the door and had whined a couple of times, but I was so tired, it didn't even register that I forgot her buttons. Standing next to the door was Stella's best chance at making me notice she needed to go outside. As much as Stella had learned to communicate with words, Jake and I had become used to Stella telling us exactly what she wanted or needed. Normally we didn't have to be on the lookout for Stella's signs that she needed to go to the bathroom. No matter where we were in the house, we could hear if Stella said "outside."

"Sorry, Stella girl, let's go outside and get your buttons." As soon as I laid out Stella's buttons in the most consistent way I could, she walked over and pushed each one lined up against the wall. She cocked her head as she heard each word.

"I think she's checking to see which one is which," I said.

At home, Stella did not need to test out all the buzzers because she learned the location of each word. A few minutes later, Stella said "water," which alerted us that she had run out.

Stella's buttons were important to her. Now two nights in a row, she became calmer and more settled after we set them up. Tonight, she even showed that she wanted to know which one was which. Her words were consistent parts of the way she communicated. Having everything I needed to express myself would make me feel more comfortable too. I thought back to a few of my AAC clients in Omaha I had observed in their classrooms at school. Oftentimes, their communication devices were not out on their desks or were not being used throughout the day. How did those kids feel when a significant part of their communication was just gone? Stella was acting out when we did not have all her tools available for her. Imagine how frustrating it would be for kids. It had not hit me yet that these clients were not mine to think about or worry about anymore. I hoped they were in good hands. I missed seeing them already.

We split up the last day of our drive and booked a last-minute room in Las Vegas. I had never been to Vegas before, so I thought it would be a fun final night of our trip. I imagined us relaxing in a nice hotel room, playing a slot machine or two, having a good meal, drinking a fun cocktail. It would be a great way to celebrate our trip thus far and pause before the chaos of moving in began.

Our hotel on the Las Vegas Strip was even more enormous than I had imagined. There were multiple restaurants inside, different wings of the hotel that all required a separate elevator, and walkways that connected to other hotels. It felt like its own world.

We walked for almost ten minutes through the main floor to reach the center of the hotel to find the right elevator for our room. Walking adjacent to the casino floor, filled with people everywhere we looked, Stella starting pooping on the carpet.

"Oh my gosh, Jake!" I shouted.

"Stella, no! Wait!" he said.

It was too late to stop it from coming out. Jake and I panicked. Jake scooped Stella up from the ground and started running away. "We gotta find the exit before she goes again!" he shouted. I ran behind Jake, with water from my fish's mason jar splashing out like crazy and Stella's buttons all shouting, "bye love you walk help eat" at the same time as they bumped into one another. I flagged down a maintenance worker on the way to help clean it up. Stella looked back to me over Jake's shoulder, smiling, like she was enjoying the thrill of a lifetime. We were officially the people everyone was staring at in Vegas, a city filled with spectacles at every turn.

Several wrong turns and ten minutes later, we finally found a door. When we set Stella down on the turf lawn, she looked up at us and kept walking. She did not need to go to the bathroom any more than she already had in the casino.

The next morning, Stella said, "outside" twice in a row in our eleventh-story hotel room. Jake walked her down the hallway to the elevator, waited a few minutes for it to arrive, and went down to the main floor. By the time he and Stella walked out of the elevator it had been at least five minutes since Stella said, "outside, outside." This time, Jake knew where the door was. He would have to walk through the casino for about five more minutes to get to it. But, as soon as they stepped out of the elevator, and started walking down the hall, Stella started pooping again. Outside was too far away for her puppy self.

At noon, we arrived at our new apartment in sunny San Diego. The wide street lined with palm trees on each side gave us a warm Southern California greeting. We were a ten-minute drive from downtown and from the beach. A small grocery store and a library sat on the end of our block. Bustling streets filled with boutique shops, restaurants, breweries, and bars were all within a half-mile walk. Lizards scattered into the bushes with each step we took. Hummingbirds buzzed past us. Succulents the size of trees grew in front lawns everywhere I looked. Green-and-red parrots squawked as they flew down the street. It felt like we had entered a completely different world.

Our apartment complex was an off-white building with sixteen units, all facing a courtyard in the center. Our unit was on the first floor, right next to the entrance. We walked in to see the refrigerator and stove in the middle of the living room, and two handymen working in the kitchen.

"Wow . . ." Jake said. "This is really . . ."

I saw the panic in Jake's face. "I'm sure it'll look bigger when the appliances are out of the living room," I said. "The listing said it was seven hundred square feet . . ."

"We'll be done in an hour," one of the workers said. "Come back then."

We walked around the neighborhood and stopped to eat lunch. I was shocked at how many restaurants were dog friendly. Back in Omaha, we could bring Stella with us to eat at only a couple spots. Stella trotted up to everyone we passed on the sidewalk, wagging her tail, and smiling. If someone said "hi" to us, but did not look down to Stella, she jumped up in the air to try catching their attention. Stella loved when people acknowledged her individually.

When we returned to our apartment an hour later, seeing the appliances placed back into the kitchen did not make too much of a difference. The long, narrow living room opened to a tiny U-shaped kitchen on the left, and the bathroom and bedroom on the right. We could see pretty much the entire place from the front door. Jake measured the whole space out to be 470 square feet.

"Well, it's a good thing we didn't bring much from Omaha," he said. "Absolutely none of our furniture would have fit." Our sectional would have taken up the entire living room. The old dining table would have blocked the walkway from the kitchen to the bedroom.

Stella ran in and out of each room, sniffing every inch of the floor. Before we started unpacking the car, we set her buttons out for her in the living room and tossed a ball across the apartment a few times so she could run a bit. When Jake began carrying boxes inside, Stella stood in the middle of the living room, smiling in the patch of sunlight that poured in through the southern-facing window. She wagged her tail every time we brought our belongings from Omaha inside. She sniffed and licked each box.

"Yes, Stella! This is our new home, girl."

I thought we had made it through the most stressful period of our cross-country move. I figured the process of finding new jobs, shedding our belongings, and traveling thirty hours in the car would be the uphill stretch for Jake, Stella, and me. I was certain that when we arrived, it would be a downhill ride the rest of the way. That was before I realized it would take Stella far longer to adjust to a new home than I expected.

Stella had used her buttons in the various hotel rooms on the trip, so I assumed she would continue using them when we set them out in our new apartment. But besides saying "outside," she did not say anything for the first few days. She was uncomfortable and filled with anxiety being in our unfamiliar apartment. She whined through the night and every time Jake and I stepped out of the apartment, even when we were just carrying the trash out. We tried leaving Stella at home while we went to shop for household necessities. As soon as we walked out the door, we could hear Stella yelping from inside. What if she didn't stop? What if the neighbors complained? The two of us could not leave at the same time anymore without her. When I went to the bathroom, Stella pawed at the door and whined. She had never been this anxious or attached before, not even as a young puppy. I wondered, *Did we completely traumatize her by putting her through this move?*

When I saw how uneasy Stella was, it did not surprise me that she was not talking. Her basic needs of feeling secure and comfortable were not being met. Before anyone can reach their fullest potential, their primary needs have to be attended to first. The noted psychologist Abraham Maslow proposed his hierarchy of needs theory to explain that humans have five different levels of needs. The different levels are illustrated in a pyramid with physiological needs—including food, shelter, and rest—at the foundation; security and safety, second; followed by social relationships and belonging, third; then feeling accomplishment, fourth; and finally, achieving one's full potential, fifth.[27] It is difficult to experience the feelings at the top of the pyramid when the foundation is unstable. We have all experienced varying degrees of this concept. It is harder to learn if you are exhausted from a poor night of sleep, and it is more challenging to focus in class when your stomach is growling. We all do our best when we feel our best. Stella was no different.

TAKEAWAYS FOR TEACHING *YOUR* DOG

- **Bring your dog's buttons, whenever possible, while traveling.** Having all forms of communication available may help your dog feel more at ease in new situations and unfamiliar places.
- **Look for your dog exploring her words on her own.** The more your dog explores her buttons and hears which button says which word, the better for her learning.
- **Keep Maslow's hierarchy of needs in mind.** If your dog isn't feeling safe, secure, healthy, or well-exercised, it will likely be harder for her to learn new skills and use them. Tend to your dog's foundational needs first.

CHAPTER 11

Using Words in New Ways

After almost a week of Jake and I alternating leaving, creating new routines for Stella, and setting up our apartment, she settled down. One afternoon she walked to the floor space under the built-in shelves in our living room and pawed each of her four buttons, "play" "no," "love you," and "help." She cocked her head to the left and right, like she had in our hotel room, listening to each word. When she heard them all, she walked away and lay down.

"Good girl, Stella. Yes, those are your words," I said. I modeled each of them again a few times in a row. I walked through the rest of the apartment to show her where "walk," "bye," "come," "eat," and "water" were. I came back to say "love you" with her button, then kissed Stella's head and patted her back. "Everything will be okay, girl."

The next day, Stella stared at the same row of her buttons again. She looked back to me, then looked down at the buttons. She timidly lifted her paw, said "no," then immediately turned her head. It looked like that was not what she was expecting to hear. Stella paused for a few seconds, then hit the button next to "no."

"Play," she said. When Stella heard that the button said "play," she pushed it again two more times. She ran away and shoved her nose into her toy bin across the room.

Stella corrected herself when she said the wrong word. When she reached the word she intended to say, "play," she pushed it two more times, like she was letting me know that was really the one she meant. Stella was now developing such strong pragmatic skills along with her vocabulary. Knowing how to use words is much more than understanding their meanings. It is equally as important to know when to use words and how to convey a message with them. I never directly taught her to push a button twice in a row to let me know that was what she really wanted. If I stepped on the wrong button, I might have modeled that idea once or twice by pushing the button I meant to say a couple of times in a row. So either Stella learned that strategy from such limited exposure, or she realized how to emphasize her words on her own.

Later that day, Stella said "water" when her bowl was empty, and "eat" within a half an hour of her normal dinnertime. She initiated play with us more and stayed by herself in the living room for a few minutes if we went into the kitchen or bedroom. Little by little, she was becoming more at ease and secure. Thankfully, I could see glimpses of the old Stella coming back.

Once we settled into our new space and saw that all our

belongings would indeed fit, our small living quarters did not bother me at all. It was clutter-free. Every object had its own place, and cleaning our entire space only took twenty minutes. Jake and I picked out our furniture together and chose not to buy a TV, which all meant that we were experiencing a new way of living together. We had exactly what we needed—no more, no less. Finally, for the first time in over a month, the three of us could fully relax. We made it. We were here, in California, ready to have fun and begin our new lives.

In Omaha, Stella said "outside," "eat," and "walk," consistently every day. Then, she would also typically use one or two other words throughout the course of the day. Now, Stella said six or seven different words each day and used many of them multiple times. She said "outside," "walk," and "eat," each morning and evening, "come" when Jake and I were making dinner and she wanted us to play in the living room with her, "help" when the blinds were closed or the bedroom door was shut, "no" when we told her we were all done playing or were "all done outside," and "bye" right after Jake or I walked out the door to go to the store. She had not only matched her former communication patterns, now she exceeded them.

Stella used her words more now that I modeled them whenever I could. This progress aligns with what I teach, and what the research shows. The more often speech therapists, parents, teachers, siblings, and peers use a child's communication device, the more the child will use it.[28] Now that all the rooms in our apartment were only a few steps away, we were always near her words. Back in Omaha, Stella could only easily say words when she was on the first floor. If she was upstairs or in the basement and really needed to go to the bathroom, she would run to let us know. But otherwise, she rarely traveled across the house to tell us "help," "come," "play," or "no." And we rarely traveled across the house to model words. By the time we reached the button we were looking for, whatever was happening was usually irrelevant. I wanted to model words when they were relevant, not after the fact. Now, pretty much whenever we talked to Stella, we would come into the living room and use her buttons. It was much easier and more convenient for Stella to access them, which led to even more communication success.

Stella paid close attention to what Jake and I were doing. If she was not playing, she was usually standing a few feet away from us, turning her head toward each of us when we talked or walked away. She was a careful observer of us and her environment. One evening after Stella finished eating, she stood between the kitchen and our dining table, watching us carry plates, glasses, and our food to the table. Jake and I sat down.

"Eat," she said and walked away toward the couch.

"Didn't she already eat?" Jake said.

"Yeah, I think she's talking about what we're about to do. Yes, Stella. Jake Christina eat."

Stella listened to my words from the couch and turned her head to the side. She lay there watching us eat for the rest of our meal. At this point, we had never fed Stella food from the table. She did not beg from us or expect us to give her scraps. She would typically sit on the couch or lie on the floor and observe. Tonight, she used the word *eat* to narrate what was happening instead of requesting food. If Stella was wanting more to eat, she likely would have stayed by her bowl or walked to the cabinet where we kept her food and treats. But she did not do that. She continued watching us.

Hearing Stella say "eat" in this new context was really similar to when she said "water" while she watched me tend to my plants. Words can be used for multiple different functions of communication and for several different meanings. The more we remember that words have multiple meanings, the more we can understand emerging communicators' messages.

The next morning at work, I was reminded of this concept again. I was enjoying my new job in early intervention so far. It was an adjustment for the first couple of weeks, as I drove to each appointment and conducted therapy in other people's homes. But I was getting the hang of it now, feeling more at ease each day. I was empowered knowing that my sessions were most families' introduction into the therapy world. It was an incredible opportunity to show them how fun, powerful, and important speech therapy is.

This morning, a curly-haired toddler shouted, "Monkey." He jumped up and down and clapped his hands in front of my face. He and I were surrounded by toys on his living room floor—a train set, blocks, Mr. Potato Head. His mom bounced his baby sister on her hip in the kitchen, within earshot of the therapy. Since starting speech therapy with this boy a few weeks before, this was only the fourth or fifth word I heard him say.

Last week I brought a couple of stuffed monkeys to play with. I left them in my trunk this week and chose other animals instead. *Maybe he wants those again? Or wait, does he want me to sing "Five Little Monkeys"?*

"Sing monkey song?" I asked. "Five little monkeys jumpin' on the bed . . ."

"Oh, no, he's been saying monkey all day long," his mom said. "He's trying to tell you that we went to the zoo yesterday and saw the monkeys. He kept jumping up and down and pointing at them."

"Oh, monkeys at the zoo. Wow!" He smiled again and resumed jumping up and down.

When toddlers are developing language, they often use single words or short phrases to convey entire thoughts. That is a completely normal part of the language development process. Stella also used the same single words to convey multiple different thoughts. She could say "eat" to request food or to comment on us eating. My client could say "monkey" to mean that he saw monkeys at the zoo yesterday or that he wanted to play with toy monkeys. Staying open to possibilities and using all the available context clues helps us figure out the speaker's intention behind each word.

A few weeks into living in San Diego, we threw together a housewarming party. Jake and I hosted my friend Brissa who was visiting us for the weekend, a fun couple we met at a brewery in our neighborhood, a couple of Jake's old friends from high school who happened to live in the area, and a couple of Jake's coworkers. It was a tight squeeze in our small apartment.

"So what are these . . . buttons you have all around your floor?" Jake's friend asked.

Another friend said he was curious, too.

Everyone stared at me, eager for an explanation. Our friends in Omaha had heard me talking about teaching Stella to say words from the very beginning of my pursuit. They understood me and my job and were not crazy surprised to see me trying to combine my two worlds together. But now here I was, reminded that lining a living room floor with talking buttons is not something people see every day.

"Well, so, I'm a speech therapist . . . and I wanted to see if Stella could learn to say words with these buttons."

Silence.

"I work with a lot of kids who are nonverbal and use communication devices to talk. So I wondered if Stella could learn to use a communication device if she had one."

Brissa caught my eye and gave me an encouraging smile. "They each say a different word," she chimed in.

"Right," I said. "Like, outside, play, walk, eat." I walked around the room, stepping on all the buttons to give a demonstration. "So when Stella wants any of these things, she can tell us."

I braced myself for all kinds of questions and skepticism. But I did not receive any.

"Wait, this is so cool."

"That makes so much sense."

"My dog totally understands words and just waits for me to say them."

I sighed with relief. I was happy to find out that I was not coming across as completely crazy to these strangers and acquaintances. My first impression on our new California friends was not shot. They were all so intrigued.

A couple hours into the party, Stella woke up from her nap, said "play," and trotted into the circle showing off her toy to all these new friends.

The next day, we were about to leave to take a walk on the beach. Brissa went into our bedroom to change. Jake grabbed my hand, turned me toward him, and kissed me.

"Love you," Stella said. We looked down to see Stella

squeezing right in between our legs. She wagged her tail, looked up at us, with her ears turned straight back against her head. Her ears always went back like this when she was happy. She looked like a little otter.

"Aw, love you, Stella. We love you, too, girl." Jake and I crouched down immediately. Stella licked our faces, smiled, and rolled over for a belly rub.

"Good girl, Stella. Good Stella, love you."

It was the first time she had ever said "love you" on her own. Any other time had been after I was modeling it for a bit while giving her belly rubs or scratches.

Stella had now officially used all ten of her words independently. I had always modeled "love you" during natural moments of connection with Stella, like a parent would with a child. Parents say "I love you" to their baby or toddler while they are hugging or kissing their child, taking care of them, or are proud of them. Parents do not explain the meaning of love to their baby first, quiz them on what it means, then accept it when the child says "I love you" to them. Children learn the social times and feelings associated with the words they hear. So far, Stella had learned in the same way. She wanted to join in on the love and had a way to tell us.

I did not worry about whether "love you" meant the exact same thing to me as it did to Stella. It's impossible to know if any single word evokes the identical feeling between multiple different people, much less humans and canines.

At the end of the weekend, Jake, Stella, and I dropped Brissa off at the airport. Stella stood in the back seat, watching out the window as Brissa disappeared into the building. When we returned to our apartment, Stella walked straight to the couch

Brissa slept on. She sniffed the blankets and pillows she used, then said "bye," and looked up directly into my eyes.

"You're right, Stella. She went bye."

This was the first time I noticed Stella talking about something that had just happened, not something about to happen or currently happening. Stella understood the concept that this person was here, but now she is not anymore. This reminded me of all the times Wrigley would sulk on my sisters' beds all day when they left for college. That was her ritual to acknowledge they had left. What was she thinking when she was lying there? And what else was Stella thinking about the fact that our friend was here for a few days and now she was gone? I did not know it then, but this would become a pattern every single time we had friends or family stay with us. Stella would ride in the car to the airport with us, sniff the air mattress or couch when we came back, and say "bye."

After work the next week, we drove west fifteen minutes to the Ocean Beach Dog Beach. Every time we pulled into the parking lot, it felt like we had truly entered a new world. Surfers adjusted their wetsuits in the parking lot, then ran off to catch the next wave. People sold art out of their vans they had turned into homes. Skateboarders zoomed past us. And on the right, there was a football field length of sand leading up to the shore where dogs of all sizes ran and played free as could be. I unclipped Stella's leash when our toes touched the shore. Stella ran up to greet every human who crossed her path. Everyone fawned over her.

Stella barked at older dogs who lacked the care or energy to keep up with her seven-month-old self. She dipped her paws in the water and bolted back to the sand when a wave came crashing in toward her.

When I threw Stella's ball down the shore, she glanced at it, then ran in the opposite direction to chase a dog who was chasing another dog. She could not care less about her ball now.

"We're at the beach, Stella! Beach, beach, beach," I kept saying as she played. "Play at the beach!" Stella smiled and sprinted past me.

The sun started to set, and the palm tree silhouettes against the deep purple sky looked like an image from a postcard. I still could not believe we lived here.

We returned from the beach feeling relaxed and happy that this could be a typical weeknight after work. I glanced at the clock to notice that it was just after 7:30.

"Can you believe this?" I said to Jake. "I would have just gotten done working in Omaha." Now it felt like I had a life. Since the end of my last appointment of the day I had already walked Stella, spent time modeling her words, made dinner, gone to the beach, and come back. It was a balanced life, not work work work.

When we finished drying Stella off, she marched inside straight to her dishes. She slammed her paw to say "eat." She turned the corner, walked to her row of buttons in the living room, and said "no." Stella looked up to us. She maintained eye contact, whined, and stomped her right paw. We did not feed Stella before we left for the beach that night. And Stella created her first two-word combination to tell us that she had not eaten dinner.

TAKEAWAYS FOR TEACHING *YOUR* DOG

- **Consider how easy your button setup is to access.** Are all your dog's buttons spread throughout the house? Are her words in a room that you don't spend much time in? If either of these is the case, consider setting up your buttons in an area that will lead to more modeling, and easier access for your dog.
- **Be present.** Limit distractions in your environment and be in the moment when you're teaching your dog. It's easier for everyone to learn in a calm environment. When you're modeling words, try keeping the TV off or turning the music down so your dog can focus.
- **Stay open to possibilities of communication.** Once again, your dog might not always be using words to request an object or action. Spending time getting to know your dog's communication patterns will help you determine when she is using words in unique ways and what else she might be trying to say. Use context clues such as what's going on in the environment, your dog's typical routines, and her gestures/vocalizations to help you.
- **Don't worry about exact translations of abstract concepts like "love you."** It is impossible to know if two humans even feel the same way when saying the same words. Your dog will likely learn to say these concepts in the same types of contexts that she sees and hears them modeled in.

CHAPTER 12

Creative Combinations

Was that a crazy coincidence? Did Stella really mean to put two words together to create a phrase? It was just too unreal. Maybe it was a fluke. Maybe it would never happen again. I texted Grace immediately, Stella said "eat no" when she hadn't eaten dinner way past her normal mealtime . . . so intrigued to see if this happens again . . .

It was not a fluke. The next morning, Stella jumped off our bed before us, like usual, and walked into the living room. I lay in bed, patiently anticipating whether I would hear Stella say she wanted to eat breakfast first or go outside first. She always told us what she wanted to do.

"Come," Stella said. Her collar tags jingled. She must be walking somewhere. Where was she going?

"Outside," she said.

I flung the covers off. "Come outside? I'm coming, Stella." Stella stood by the front door wagging her tail. She went to the bathroom as soon as we reached the courtyard.

Back when I had introduced a few buttons to Stella, I remember joking around with Grace about how crazy it would be if Stella ever said two words in a row, like "play outside." We laughed, and I never seriously thought about that wild scenario again. It was the same sort of hypothetical situation as saying, "How cool would it be if we won the lottery tomorrow?" I had never modeled two words together with Stella's buttons. It never seriously crossed my mind that she might reach this significant language milestone. I modeled two-word phrases all the time verbally, unintentionally—"come girl, come outside," "come eat," "Stella no," "play toy." But I was only doing that because it's natural to talk in short phrases to help her understanding, not because I expected Stella to follow my lead.

The more I thought about Stella's new skills, the more it actually made sense. If she was already using single words for multiple meanings in several contexts, why wouldn't she be able to use them together to create new phrases? Toddlers combine words together when they have a solid understanding of their meanings and have practiced using them individually first. So far, Stella had continued sharing language milestones with toddlers. Why would the similarities stop now? Plus, Stella had been combining single words with gestures regularly. She would say "help" then stand where she wanted us to look for her toy, "walk" then paw at the door. Pairing words and gestures occurs before kids use words to say both concepts. Toddlers typically start combining two words together when they are about eighteen months old. Some skills that happen developmentally right before we see word combinations in children include:[29]

- "Uses single words frequently." Stella communicated with single words several times each day.
- "Verbalizes two different needs." Stella used her buttons to tell us when she needed food, water, and to go to the bathroom.
- "Uses words to interact with others." Stella engaged with us by using gestures, vocalizations, and words.
- "Understands the commands 'sit down' and 'come here.'" Stella responded to these, along with a few other commands.
- "Requests assistance from an adult." Stella said "help" when she needed it.
- "Talks rather than uses gestures." Stella used words for the concepts she had available to her.
- "Uses words to protest." Stella said "no" when we were doing something she did not like.

Even with such a limited vocabulary available to her, Stella was still using words in similar ways as toddlers do before they start putting separate words together. For the first couple weeks of combining words, Stella's most common phrases were actions paired with "no" if we had not done something that we usually do.

One evening, I came home from work, wiped out from a few chaotic sessions. Instead of taking Stella outside to go for a walk right away, I flopped on our bed, hoping to rest for a few minutes. Stella walked into the living room, said "walk no," and poked her head into the bedroom. She looked concerned.

"We'll walk later, Stella. Come up." I patted the bed.

Now, by combining words with "no," Stella could talk about what she noticed in even greater detail. This gave me even more insight into how routine-oriented Stella is. I knew she thrived with structure and anticipated what was coming next, but Stella pointing out when something that usually happened had not occurred yet showed me that she was thinking about it even in its absence. What else was she thinking about that was not happening right in the moment? A few minutes later, Stella hopped off the bed.

"Come walk," I heard from the living room.

I could not resist. I would rest later. First Stella pointed out that we were late for our normal walk, and now she is requesting it. "Okay, girl, let's go for a walk." Stella jumped around in a circle then stood by her leash.

Taking Stella to the dog beach quickly became a routine activity for us. A few times per week I would either take Stella when I came home from work, or the three of us would go together after dinner to watch the sunset. Running with Stella down the shore, meeting new dog and human friends, and enjoying the beauty of the Pacific Ocean was the perfect way to end the day.

"Do you want to make a quick dinner then head to the beach?" I asked Jake.

I heard Stella's tags jingle immediately. She pranced into the bedroom and cocked her head up at us.

"She definitely knows that word now," I said.

"Yes, let's go to the beach," Jake said.

Stella turned her head again, then smiled until we put her leash on. As soon as we pulled into the dog beach parking lot, Stella perked up from the back seat of the car. She looked out the window and started whining and wagging her tail when she saw dogs trotting around.

The night Jake and I picked up Stella (in a grocery store parking lot, of all places). We had no idea how our lives would change, and neither did she! *(Right)* Little Stella staring into my eyes minutes after bringing her home. I instantly felt connected to her.

Stella studying her "outside" button three weeks after I introduced it. Based on how frequently she stared at it, I could tell she was close to pushing it on her own.

After Stella started saying her first three words independently, I was hooked on stories of human and animal communication. I loved when Stella would keep me company while I read!

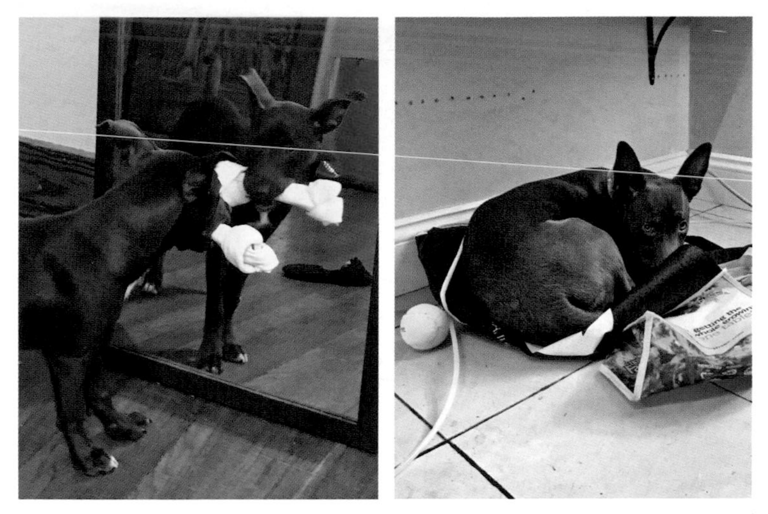

Whenever we gave Stella a new toy or bone, she would go stare at herself in the mirror. I always wondered what she was thinking when she saw her reflection. A few minutes later, she buried that bone in a pile of Jake's clothes. *(Right)* Stella's body language and word use showed us how stressed she was while we packed up our Omaha house. After we removed the last of our couches, Stella said "no" and laid on an empty grocery bag.

On our road trip to California, Stella enjoyed taking in the beautiful new scenery just as much as we did.

About a week after we arrived in San Diego, Stella started using her words consistently again. She told us "eat" every evening without fail at five o'clock on the dot.

Stella adapted well to her new California lifestyle. The dog beach quickly became one of Stella's favorite places, and "beach" became one of her most frequently used words.

Stella looked happy and excited when we responded to her words. She waited for me by the door after she said "walk" and I said "yes."

The first version of Stella's buttons placed together on one board. Once she adjusted to her new setup, Stella loved hanging out on and near her board.

Her left paw on her ball and her right paw saying "play." When Stella's words became automatic for her, she often said them while multitasking.

Stella with two of her best dog friends at the park. She initiated the same games with the same dogs each visit. That reminded me of toddlers initiating the same games I played with them every week.

Stella was a fast learner! Shortly after we switched her buttons to a single board, she was ready for new vocabulary. Every day, Stella generated new word combinations spontaneously.

I thought fifteen words would keep Stella busy for a while, but I was wrong! We transitioned to a larger board with more vocabulary when she used all of her words appropriately and consistently. Stella laid on her old board for a full afternoon before she tried using her new one.

Since we kept all of Stella's buttons in the same order, the switch to a larger board was much easier than when we first grouped her buttons together. Stella's device now took up a big chunk of our living room!

It fascinated me how often Stella would comment "park play" or "park happy" after we returned from the park. I wondered what else she was thinking about her trip, and what more she could say about it if she had the vocabulary available.

Even though she uses words to talk to us, Stella still acts like a dog, and communicates like a dog. The only difference is that she has one more tool to express herself.

With all of her buttons securely fastened to the board, we can travel with it and bring it places without having to set it up every time. Stella always enjoys having her words with her.

I knew what this meant. Stella needed a "beach" button. After she heard the word *beach*, she always tracked our every move and tried herding us out the door as quickly as possible. If she was understanding "beach," she should definitely have a way to let us know that is what she wants instead of waiting for one of us to mention it.

"Ready, girl? Let's play on the beach!" I said.

The next day, I found our last spare button in a cabinet. While I put batteries in it, Stella stood close to my side and stared up at me. She knew what these buttons were for by now. I held it out for her. Stella sniffed it, then watched intently, as I said "beach" into the speaker. Stella barely waited for me to set it down near the front door before she pawed at it repeatedly.

"Beach beach beach," Stella said.

I had a feeling something like this would happen, just like when we added a "walk" button for her. Stella could not wait another second to say one of her favorite words.

"Okay, Stella. Let's go to the beach," I said. Luckily I programmed the button when I could actually take her. I am sure Stella would have been so disappointed if I had to say "no" the first time she had the chance to tell me she wanted to visit her favorite spot.

A couple of nights later, I woke up to Stella saying "beach, beach" at 3:00 in the morning. Stella rarely talked in the middle of the night. Once or twice she told us "outside" when she really needed to go to the bathroom. And once last month, she said "help" in the middle of the night, which startled me. It was 2:00 A.M., and when I turned on the light, Stella was chasing a cricket in our living room. But other than those instances, she had never woken up to talk in the middle of the night.

"No beach, Stella. Bed now."

It was not always possible to take Stella to the beach exactly when she asked for it. But it's okay to say no. Stella already had a solid understanding of the word *beach*, from us saying it every time we took her there. In AAC therapy, "once the individual has the idea of what a particular word means and understands that there will be a natural response when appropriate, then it's okay to respond by saying 'no more right now,' 'we're finished with that,'" or something else of that nature.[30] Saying no to a request can still provide reinforcement for the learner. For example, if a child asks for cookies all day long, you would at some point probably tell them "it's not time for cookies now," or "we can have cookies later." You would not be afraid that by saying no, the child would lose understanding of the word *cookies*. You still acknowledged and responded to what they said, which provides reinforcement for their learning.

A lot of parents and professionals can be afraid to program highly motivating words into a child's device. I have heard the same comment so many times, "He'll just ask for it over and over again, so I don't want to start that." I am always thinking to myself, *Isn't that the point of communication? To be able to talk about what you want to talk about? And to communicate what's on your mind?* Imagine someone saying the opposite statement, "He'll never want to say this word so we should definitely have it available for him." We would never choose that because it does not make sense.

Even if it is inconvenient for us for a little while as they are learning, everyone deserves to be able to say the words they want to say *when* they want to say them. Communication should not be conditional. I have seen many adults remove words from a child's device or put the child's talker away because they were becoming annoyed by how often the child said something. This would be the equivalent of duct-taping a child's mouth shut if they kept asking for "cookies" over and over again. Instead of removing the word from the device, or hiding the device altogether, it is best to respond by saying "no," "all done," or "later." This gives the AAC user a chance to learn boundaries and understand the meanings of those words, too. It can also provide a good model for how the AAC user could say no to something that they do not want. The point of teaching language is not to hear what we want to hear when we want to hear it. We teach words to empower others to share their own thoughts, whatever and whenever they may be.

With time, practice, and patience, children learn when it is a good idea to ask for something, and when it is not. It is a learning process. Stella did not continue saying "beach" ten times each day forever. Eventually, she learned the patterns of when we typically go and requested "beach" during more appropriate situations.

When Daylight Saving time arrived in November, we turned our clocks back one hour. Stella showed us, once again, how routine based she is. Between 3:30 and 4:00 P.M., she repeatedly requested to "eat." This would have been 4:30 or 5:00 before the time change, which was completely normal for her. But I did not want to feed her dinner so early and throw her off for our workweek ahead. I gave Stella a couple of treats to tide her over but kept saying, "No eat now, eat later."

Fifteen minutes passed.

"Help eat," Stella said then barked.

"I know, Stella, good waiting. Eat later."

Stella sighed. She stood still for about ten seconds.

"Love you, no," she said. Stella walked away, into the bedroom.

Jake's jaw dropped. "Oh my gosh . . ."

I held my hand over my mouth. "I can't believe . . ."

When we were not giving Stella the response she expected, she adjusted her message to say, "help eat." And when we still said no, she let us know that she was not happy with us. She not only combined words and modified her message so appropriately, she also engaged in one of the first actual short conversations we had. Normally, Stella told us what she wanted or what she was thinking, and we would respond. This time, she heard and saw my response that I was not going to feed her right then and replied to it by saying "love you no."

Generating novel utterances is the ultimate goal and purpose of language. According to the American Speech and Hearing Association, "The vast majority of the sentences we use in our daily communication are sentences that we have never used before in our lifetimes. Furthermore, those *sentences have never been spoken by anyone in the history of mankind.*"[31] When we teach single words first, the possibilities for communication are endless.

I had never modeled "help eat" or "love you no." Now Stella was past the point of using words on her own that she saw us model before. She was past the point of combining two words to say a phrase she had heard us verbally say before. She was stepping out to create her own messages and use words in ways unique to Stella. She took pieces of what she had heard and learned and put them together to create something new. This was always one of my favorite parts of teaching language. Seeing a child progress from saying mostly single words that they have heard before to suddenly putting them together to create their own expressions seems miraculous every single time. And now, I was witnessing this with my dog. *What is Stella's actual potential here?* I wondered. *Have I even scratched the surface of what she is trying to communicate to me or what she can learn to say if she has more vocabulary available?*

By mid-November, Stella was generating at least one new two-word phrase each day. Even with only eleven words available, there were still 110 unique two-word combinations possible that she could make. She also used a few of the same phrases throughout daily activities. Some of her frequent messages included "come outside" when she called us over from the bedroom to go out, "bye walk" when she stood by the door ready to leave for her walk, and "love you play" after we spent a longer time than usual playing with her inside. While I did not initially expect Stella to combine words, the words she had available enabled her to reach this milestone. When I selected her vocabulary, I picked mostly core words that I knew she understood and could be used in multiple contexts. If I had selected primarily nouns, Stella would not have had the opportunity to create such functional messages. Verbs are necessary to create phrases. We say phrases like "play outside," not "stick ball." We say, "eat dinner," not "food treat." Providing a solid vocabulary from the beginning paved the way for Stella to achieve these more complex language milestones.

It only took Stella about one month after settling in her new home to explode with language use and combine words. I wondered how quickly she would have reached this breakthrough if her buttons were spread out in one room rather than across the first floor of a house, or if she did not need time to settle into an entirely new space. Stella had been through a lot over the past couple of months. But, even with all these changes, she was still learning so quickly. She was only nine months old and she talked to us in short phrases every day.

One day while I was washing dishes, I looked out the window to see Jake and Stella returning from their walk. She carried a five-foot-long palm branch in her mouth, looking like the proudest dog in the world. She loved picking up giant palm branches on walks. She shredded them and frolicked with a piece of it for a block or two until she lost interest. Jake let her bring it through the gate, then had her drop it in the courtyard before coming inside.

"She's obsessed with this one," he said. "She wouldn't leave it behind."

Stella took a big drink of water and smiled. She turned the corner to say "play," then walked over to the door to say, "outside."

"You want to play outside, Stella?" I said. I put Stella's leash back on and took her to our courtyard. She immediately pounced on the palm branch, bit it, and shook it back and forth.

She could have only told me "outside" to say that she wanted to go back out. But she said more than that. She combined "play" and "outside" to let us know specifically what she wanted to do. She did not have to go to the bathroom, she did not want to go for another walk, she wanted to keep playing with the prized palm branch that she carried all the way home.

Stella's motivation to share her thoughts with us must have been inherently strong. It was not even easy for her to put words together. She had to push one button, walk across the room or around the corner, then push another. Ease of access to communication devices impacts use. I saw it firsthand with

Stella. Living in our apartment with all her buttons always only steps away was easier for her compared to living in a large house with her buttons spread everywhere. And I witnessed it at work. Some AAC systems require children to navigate through six or seven pages to find specific vocabulary words. When kids use devices that allow them to find every single word by pressing only two or three different icons, it is much easier for them to talk. I wondered how I could make this easier for Stella. Would she communicate even more if her buttons were all in the same place? Or would it be too difficult for her to differentiate between them?

I did not want her to have to walk across the apartment to communicate her entire thought to us. I briefly thought about moving all her buttons close to one another. Maybe I could dedicate one part of the living room to all her buttons. But we were leaving to go back to the Midwest for the holidays in a few weeks. It was not the right time to make any sort of big changes to her button setup. I wanted Stella to be able to communicate with ease while we were gone, and I wanted to be with her to see how she reacted to a new setup. This way I could support her learning and decide if I needed to make adjustments.

The week of Thanksgiving, I caught a terrible bug. The fever and loss of my voice kept me home in bed for an entire week. Stella lay right by me each day, all day long. She curled up in a ball and rested her head on my chest. Every so often she licked my face. This was strange for Stella. She never lay up on the top half of the bed, she almost always stayed near our feet.

On Monday, the worst day of my sickness, Stella did not

request anything from me. She did not ask to go outside or to play, she did not ask me to take her for a walk or go to the beach. She knew I was not going anywhere. She only hopped off the bed once in the middle of the day. I listened while she walked into the living room.

"Love you," she said.

Stella ran back into the bedroom and returned to my side.

"Thank you, Stella. I love you, too, girl."

Later that week, after I finished watching the last Netflix documentary to capture my interest, I sat up on my bed, and looked around at the blank walls. I had read all my books from the library, had watched everything remotely fascinating, and had no idea what to do for the rest of the day. I stared at Stella. *People need to know what she can say*, I thought. *What would happen if the world knew what she could do?*

I opened a blank document on my laptop. I had no idea what I was doing and had no plan for what I was going to write. But somehow it felt like my fingers knew exactly where to start.

Teaching My Dog to Talk, I typed at the top of the page. Looking at those words made me smile.

I had no idea who I was writing this for, or what I would do with it someday, or if I would even finish whatever I was starting. But thoughts were coming to me from all directions. I was remembering what I learned in graduate school, stories from being a speech therapist, reflecting on all that I had seen Stella accomplish so far. I wrote out questions that continuously popped into my head, and ideas for what more I could try with Stella. I typed and typed and typed.

Before I knew it, I had a new folder on my desktop filled with documents of ideas, questions, and explanations of our journey so far. Maybe someday I would write a really cool article about this experience that I could submit to a magazine. Or maybe I could be featured on someone's website or blog. All I knew was how excited I was to be reflecting on Stella's communication skills and thinking about how to share this story when the time was right.

TAKEAWAYS FOR TEACHING *YOUR* DOG

- **Model two-word phrases.** Help your dog learn to combine words by talking in short phrases, and using your dog's buttons as you talk. When your dog starts using single words frequently, be on the lookout for word combinations.
- **Program words that your dog reacts strongly to.** Are there any words that your dog waits around for you to say, or that you have to spell out so your dog doesn't overhear your plans? If your dog understands a word, give her a chance to be able to say it too!
- **Keep all words available for your dog to say.** If you can't say yes to what your dog is asking for, or if your dog is asking for the same thing over and over again, respond with "no," "all done," or "later" instead of taking your dog's buttons away. Give your dog a chance to learn boundaries, and how frequently to ask for something. It takes time to learn the social rules of language.

CHAPTER 13

Help!

The morning after we returned from Christmas in the Midwest, Jake and I prepared for a hike. As we packed our backpacks, Stella paced across the living room.

"Come come come come love you," Stella said. Her body language completely changed from being so happy since we reunited with her yesterday, to a concerned, nervous puppy, hoping her owners were not leaving her again. Her tail hid between her legs and she slouched next to the door. She was telling us she wanted to be able to come with us this time.

"Yes, Stella come! I love you, too. We're going for a hike. Jake, Christina, Stella walk."

Stella's tail wagged as I clipped her portable water dish onto my backpack. I scratched behind her ears and gave her a kiss. "We're not leaving you, don't worry, Stella girl."

I had missed her, too. We had not left her for more than a long weekend before. Jake and I logged into the doggy day care webcam every day while we were away to check and see how she was doing. It looked like she was having a blast playing with her dog friends, but I wondered if she was ever worried that we were not coming back. *Did she think about us or her home at all while we were away? Did she think she lived at day care now? Or did she just think about playing with all her new friends?*

Jake and I had spent the flight to San Diego brainstorming how we could arrange Stella's buttons to keep them all in one location. "What if I made a giant AAC device for her? I could get a poster board, set all the buttons down in rows. It would look kind of like the devices I use at work. And we could keep enough space in between the buttons for her to walk through them," I said. If all the buttons were on one board, we could move it to different rooms, or easily bring it with us when we traveled. And Stella would not have to relearn the button locations everywhere we went. They could actually always remain in the same place.

Changing AAC devices can be a challenging time. I have introduced multiple new setups to children, and each one of them had a different reaction. Some kids are intrigued by the new vocabulary available to them right away and start exploring it. Some kids throw the tablet against the wall out of frustration. Some kids switch back and forth between using their old device and their new device, figuring out which one serves them best. I wondered what Stella would do.

I worried, too . . . What if Stella lost all her progress? What if she would no longer be motivated to communicate with words because I made it too challenging for her? What if I had the intention of helping, but ended up taking away her ability to talk? What if this concept only worked by keeping buttons in separate locations for her to distinguish between? It was uncharted territory. The plan seemed sound on the plane ride, but now when I was back home seeing Stella talk so well with her current setup, I did not know if it was the right choice to try to change it all on her.

As a speech therapist, I never quite understood how difficult it would be for parents to hop on board with introducing a new AAC system to their child. I could only see the immense opportunities for learning and communication that it would open up. I expected parents to be thrilled by the potential. But most of the time, they were apprehensive and tentative. They had likely spent years with their current system and had already put so much mental energy into learning it. And they could see firsthand when challenges stressed their child out at home. This is why it is crucial to start out with a great system that will support the child's language growth for years to come. With kids, excellent devices exist where we can set them up for years of development from the beginning. They should not have to undergo major AAC changes. The big difference with Stella was that I did not know what her potential could be or where we were heading. We kept progressing step by step, testing, experimenting, and learning as we went.

I thought back to Oliver and how much language and communication he had access to after we upgraded his device. I could not just sit here and watch Stella continue to progress and not adapt. I had to try something new. I had to give her the chance to thrive. As Grace reminded me, "Worst-case scenario, you can always go back. Might as well give it a shot."

A couple of days later, I picked up a half-inch-thick tan, foam poster board from the store. It seemed thick enough to not slide around or rip, but certainly light enough to carry with ease. When I came home, Jake laid the board down in the dead space on the far side of our living room, a few steps away from the front door. I walked around our apartment, picking up all nine of Stella's buttons. Two of her buttons ("walk" and "bye") broke while we were gone, but two more boxes of buzzers were on the way to us. Stella stood in the middle of the living room, watching my every movement.

The board would fit fifteen buttons while still giving her room to walk between them. I figured that would be plenty of words for a while at least. I placed the buttons down in a three-by-three grid on the left side of the board, leaving space on the right side to add more words when our new buzzers arrived. Stella stood over the board, looking down at her new setup.

"Stella, look. They're your same words," I said. I pressed each one in a row while verbally saying the word again. Stella looked up to me, then ran away. She jumped on the couch and curled up in a ball. She looked stressed out already.

"Love you, Stella," I said while I pushed her "love you" button. "It's okay, girl." Stella jumped off the couch. She lowered her head, put her tail between her legs, and walked into the bedroom. She looked scared of the changes happening out in the living room.

For the rest of the night, I modeled the words. I went back to the frequency of modeling that I started with at the very beginning, repeating each word several times verbally and with her button in its new location. Even though Stella knew and used all these words, this was a relearning period. It reminded me of when I picked up Jake's phone and tried finding an app on it. Even though we had many of the same apps and all the icons looked identical, I had no idea where they were. Jake's organizational system on his phone was completely different from mine. I usually had to ask him where to find the app I was looking for or search the name of it.

Stella watched me from a few feet away, but never came close to the board. She would retreat to her bed, the couch, or to my feet, and keep looking up to me with her big, sad puppy eyes. Her world had just changed drastically. Something she used every single day looked completely different to her now. Her body language called out for love and support. I would give this a few days, then return back to her old setup if it didn't seem to be working out.

The next morning, I walked out of the bedroom with Stella. Normally she told us what she wanted on her own, but now I knew I would need to help her out. She walked straight to her dishes and pawed at the empty space where her "eat" button used to be. I watched as she then walked into the living room, over to her board of buttons, and started pawing at all of them. Each time she activated a word, she cocked her head to the side, and pushed it one or two more times. This is such an important step for AAC users. She was spending time exploring her words to figure out how they now worked.

When she pawed the "help" button once, she continued on to say, "help help help help help help." She asked for help like I would when I could not find something on Jake's phone. Stella hopped off the board and stared up at me.

"You need help, Stella? Here, girl." I sat on the floor to be at her level. I reached my hand to the top row to model "eat" three or four times. Stella wagged her tail, licked my face, and trotted over to her food bowl.

Once again, I saw how strong Stella's intrinsic motivation to communicate was. We all have two types of motivation:

intrinsic and *extrinsic*. Intrinsic motivation is the inner drive someone has to do a specific activity. Extrinsic motivation is the desire to do something for a reward. I was not giving Stella treats or any sort of reward besides the natural response to her communication when she explored her new board setup. Even though extrinsic rewards like treats may seem beneficial, research actually shows that providing external motivation can hinder long-term intrinsic motivation in humans.[32] Recent analyses in the areas of intrinsic and extrinsic motivation has shown that, "128 experiments lead to the conclusion that tangible rewards tend to have a substantially negative effect on intrinsic motivation. When institutions, families, schools, businesses, and athletic teams, for example, focus on the short-term and opt for controlling people's behavior, they do considerable long-term damage."[33]

The entire idea of intrinsic motivation was first discovered in primates who were more motivated to solve puzzles when there was no food reward attached to it. Monkeys who explored the puzzle purely for curiosity and fun consistently outperformed the primates who were given raisins as a reward.[34] I hypothesized that dogs would also have an overpowering intrinsic drive when given the chance to use it. Treats were certainly not everything to Stella.

Stella could have given up. She could have continued pawing at her food dish and water bowl. She could have stood by the door whining whenever she wanted to go outside. She could have nudged the collar she wears when we take her to the beach to let us know she wanted the beach. But she did not do that. Her communication skills had already grown so much beyond simply requesting actions. She was motivated on her own to figure out the new locations of her words, so she could use them again. She needed her words back to share all the thoughts she had about what was happening, what she noticed, and what was missing from her routines.

She was lying on the couch, shredding her plush parrot to pieces, leaving a mess of stuffing. Stella was stressed. This was the second toy she completely destroyed in the last two days. She had not played like this since she was much younger. It looked like she was taking her frustrations out through her play.

She was whining so much more, too. Earlier, I walked into the living room to see Stella standing helplessly by herself, lost and upset. I petted her and threw a ball for her, but she wouldn't settle down.

I hated seeing her so stressed out. I thought about putting her buttons back in their original locations many times throughout those first couple of days. Maybe she would not develop more advanced skills than what she had already showed. Maybe causing her to be so stressed out would not be worth it. But I reminded myself Stella might be able to communicate to her fullest potential. *One more day*, I thought, then I would return the buttons to their original spots.

On day three of the switch, there was hope. Stella showed me that this pursuit was worth our while. She walked to her board and pushed all her buttons again. When she landed on "help," she said it four more times.

"Help help help help." Stella walked to the spot next to the door where her "outside" button used to sit. She pawed the empty space, barked at me, then said "help help" again. I walked over to her board, modeled "outside," a few times, then took her out to go to the bathroom.

Later on, when Stella finished her water, she licked the empty bowl, pawed where her "water" button used to be, then walked across the apartment.

"Help," she said.

"Stella want help?" I walked to her board. "Water, water," I said verbally and with her button. Stella watched my foot closely like a curious student. She was eager to learn where each word was.

This pattern continued with nearly every word, several times over the course of a week. Sometimes Stella would say "help" over and over again without pawing at any of the previous button locations. When this happened, I sat down next to her board with her, and modeled all her words a few times in a row. It felt like the two of us entered our own bubble of communication together.

"Help," Stella said.

I started walking over to her board.

"Love you help." Stella wagged her tail as I approached.

"Love you, too, Stella. You need more help?" I started from the bottom and pushed each button while slowly saying each word. "Love you, play, outside . . ."

Stella wagged her tail, walked to the top of her board, and said "outside" for the first time on her new setup.

"Okay, let's go outside, Stella."

At this point, Stella's language skills, her problem-solving capabilities, and the awareness she had for when she needed our assistance were extraordinary. Stella's intelligence was on full display as, even in this overwhelming situation, she found a creative way to figure out her new setup. "Help" was never one of Stella's most frequently used words. But in this unfamiliar context, she used it over and over again to learn where all her words were.

Day by day, Stella became more comfortable with her buttons again. She spent time on her own pressing different buttons to see what would happen, and she paid close attention when Jake and I modeled her words. Unsurprisingly, Stella learned her favorite words the fastest. At the end of the first week with Stella's board, I stopped home for lunch in the middle of the day when one of my appointments canceled. Stella greeted me with her typical happy self.

"Beach beach beach beach beach," she said.

Maybe she did not mean that. Maybe she was still exploring her buttons, I thought. I rarely came home over lunch, so I was not sure if it would be normal for Stella to ask for the beach in the middle of a weekday. I was still chewing my sandwich and had not yet responded to her.

"Beach beach," she said again. Stella looked back to me. She walked over to the collar she wears on the dog beach, sniffed it, and looked up at me again.

She proved me wrong, and I could not have been happier.

"No beach now, Stella. Sorry, girl. Let's go outside though," I pushed "outside" with my foot a couple of times, then took her out to the courtyard with me. Of course she would learn "beach" again quickly.

Exactly one week after we switched her setup, Stella was talking just like she was before. Her use of each word matched how she typically said it before we moved her buttons. Seeing this consistency showed me that Stella had learned all the new word locations. In addition to being back to her typical chatty self, Stella started using her board almost as a home base. She brought her toys to it, dropped them on the empty space, sat on it, and licked it frequently. Now she sometimes chose to nap lying down on the floor with her head on her board. Before moving her buttons, she never hung out in this area of the apartment. Stella was taking ownership of her device. She knew it was for her and wanted to spend time near it.

Children reach this stage with their devices, too. When kids start taking more ownership and recognizing that the talker is theirs, they carry it around with them if they can, hug it, swat other people's hands away from it. It made me happy to see Stella wanting to spend time by her board and claiming it as hers. In one short week, Stella went from looking continuously stressed out and confused, to exploring words on her own and asking for help, to returning to her same communication patterns and choosing to spend time by her board. I wondered what the next week would bring.

Stella does not cope well with rain. It rarely comes to Southern California, but when it does, Stella is beside herself. When we try taking her out for walks, she runs back toward the door. She would only go to the bathroom under the stairs in our courtyard, where she could avoid becoming too wet. Rainy days meant that Stella would be bouncing off the walls in our apartment. Without taking long walks or running on the beach, she was an unstoppable force of energy. On the third straight day of rain, we had not been on a long walk or to the beach in days.

Stella let out a sharp, high-pitched bark that reverberated off the walls in our apartment. She marched over to her board.

"Beach no," she said. She stared at me, then barked again.

"Yes, Stella, we haven't been to the beach. No beach, it's raining."

Stella whined and looked out the window.

"Come on, girl, let's play with your toy!"

Stella sulked in front of the door.

"It's okay, Stella. Let's play!"

"No," she said. She whined again, then lay down with her head between her paws.

Stella had officially surpassed her previous language use. After a solid week of relearning, she started combining words nearly every time she talked. It must have been so much easier for her to create phrases now that she did not have to walk across the room right in the middle of her thought. She talked much more often and generated even more novel utterances.

Even though Stella knew where all her words were, her hind paws sometimes activated buttons when she walked through her board. Every time this happened, Stella dramatically turned back to look at her paw and back to me. It looked like she was trying to show me she did not mean to say that word. Over the first week, Stella experienced mis-hits like these almost every time she walked through her board. But now, they were starting to decrease as she adjusted and carefully stepped through the rows without bumping into buttons along the way. When given the chance and time to learn, Stella always thrived.

On day ten of her new setup, another rainy day, Stella reached a new milestone.

"Help beach love you," Stella said.

Jake ran out of the kitchen. "Was that her?" he asked.

"It sure was," I said. Stella just said two intentional words and a phrase in a row, when we had not been to the beach in a couple of days due to rain. The only other times Stella had said three consecutive different words was when she was clearly exploring buttons. But this time was different. She navigated to each button so deliberately. And the phrase made sense for the current circumstances. It was normal for Stella to say "help" when she appeared desperate. And it was typical for Stella to use "love you" almost like "please." Stella often added "love you" to her request if we said "no" the first time.

While Jake and I were still in shock that our dog told us a three-button phrase, Stella kept talking.

"Love you water," she said. Stella stood next to her beach collar. Now Stella used "water" as another way to refer to the beach. Stella found a different way to ask for the beach when her desires still were not being met. This was an entirely new level of complexity.

Those first ten days of Stella using her new board inspired so much hope and encouragement. It furthered my belief that we have to try new things to see new results. I wondered what would have happened if I started teaching Stella words with a setup like this from the very beginning. Every time Stella reached a new level of word use, more questions and ideas arose. Would she have combined words faster? Or would it have taken her longer to learn because the words were not right next to whatever she was asking for? It was impossible to know what it would have been like if I had chosen to try this sooner, or from the start. All I could do was keep progressing forward one step at a time, observing patterns, asking questions, and brainstorming solutions. Maybe someday I would have another dog or someone else would try starting off with a full board of words to compare learning speeds. Now that Stella excelled with her new setup, I felt ready to fill up the rest of her board.

TAKEAWAYS FOR TEACHING *YOUR* DOG

- **Be prepared to support your dog if/when you need to change her device setup.** Even if your dog has been independently using words for a while, she will need help from you to reach that stage again with words in new locations.
- **Model words like crazy!** Model words as often as you did in the beginning, when you were introducing each one. The more your dog sees each word in use, the better chance she'll have to learn its new placement.
- **Make communication as easy as possible.** If you choose to keep all buttons on a single board, allow room for your dog to reach each word, walk around the board, or walk in between the rows.
- **Give your dog a chance to learn.** It can be tempting to revert to an old setup or to give up if you haven't seen progress within a couple of days. Adjusting and relearning can take time. Give it at least a week or two before reassessing.
- **Pay attention to your dog's communication patterns.** When your dog starts using words in similar ways to how she did before changing her setup, you'll know she has reached the same level and is ready to progress even further.

CHAPTER 14

Becoming Automatic

"Do you think Stella would say our names if she could?" I asked Jake. I sat on our couch staring at the two new boxes of Recordable Answer Buzzers. I already had a list of thirty potential words I was considering adding to Stella's board. But we only had room for six at the time. It was so hard for me to choose. I wanted her to be able to say all the words she possibly could. Once again, I asked myself, *What does Stella hear us say all the time? What words is she already understanding? What words will help her communicate about a variety of experiences?*

One of the first ways toddlers typically combine words is by saying a name with an action or object, like "Mommy drive," "Daddy ball."[35] Since Stella was combining words now, I wanted to give her the opportunity to create phrases that were developmentally appropriate for the skills she was demonstrating. Adding all our names could give Stella a chance to tell us who she wanted, or who she was talking to. I wondered if Stella ever thought about Jake or me when we were gone. Would she ask for one of us if we were not there? Or would she talk only about whoever was home at the time? Would she specify who she wanted to take her outside or for a walk? Would she use her own name to clarify when she wanted to do something rather than when she was commenting on what Jake and I were doing?

I recorded "Stella," "Jake," and "Christina," into three of the six new buttons.

I chose to use "Christina" and "Jake" rather than "Mom" and "Dad." Stella never heard the words *Mom* and *Dad*. Stella had been living with us for almost a year now and had heard us call each other by our names constantly. I also knew that she recognized and understood our different names. After work when I would tell her "Jake's coming," she would run to look out the window for his car. At the dog beach, Jake would tell Stella "go look for Christina." Stella would run around until she found me. We also typically narrated what we were doing with Stella and included our names. Phrases such as "Stella Jake walk" or "Stella Christina play outside" were common for us.

I replaced the "bye" and "walk" buttons that had broken a couple of weeks ago, which meant there was only room for one more word at the time. With her current vocabulary, Stella could tell us what she wanted to do, where she wanted to go, request assistance, protest "no." But she did not have a way to let us know when she liked something we were doing. "We tell her when she's being good all the time. She should have a way to tell us when we're being good in her eyes, too," I said to Jake. I programmed "good" into the sixth button.

Looking down at Stella's words, I thought back to my graduate school AAC class. We had to create communication boards with only nine or twelve words available, then actually try communicating with them instead of using verbal speech. We would know if we selected effective words if we could use the board while interacting with others. If we were repeatedly at a loss for words, we knew that we needed to make changes to it. Now, I spent time by myself testing out how to say common phrases I used on Stella's board. "Good Stella," "Stella play outside," "Christina help Stella," "Jake Christina eat," "Love you Stella," "No beach. Walk," I practiced with my right foot. So far, so good. There were so many functional phrases I could say now.

When communication partners are familiar with the locations of words on the learner's device, they are able to model vocabulary much more efficiently. By practicing now, I was already creating my motor plans so I could model quickly without scanning her board each time I talked to Stella. I learned the importance of this and how fast I could learn the locations of each word after I attended a training for the Language Acquisition through Motor Planning (LAMP) therapy approach. At the workshop, the presenters directed us to say a new word on a communication device five times in a row. The communication devices we used had thousands of words, not only fifteen like Stella's board. After we said the word a few times in a row, they prompted us to close our eyes and try saying the same word again. We could still say the word, or be centimeters off it, with our eyes closed. This was possible because we had all developed the motor plan for where the word was located. Again, this is what happens when we keep word locations in the same place; we become automatic in finding them. Motor planning is not only used in typing or saying words with communication devices. We all motor plan throughout most of our activities, and we don't even realize what our brains and bodies are doing. When we tie our shoes, shift the gears while driving, grab a fork out of the silverware tray without looking, perform a choreographed dance routine, or play an instrument, our bodies are acting automatically after learning the movements from repetition.

"Stella come play," I modeled.

Stella trotted into the living room.

"Stella Christina play," I said before throwing her ball across the room.

Now that Stella was combining two words commonly and three words occasionally, I modeled three- and four-word phrases as much as I could. I talked in these short utterances to Stella and also added on to her words when she said a one- or two-word phrase. This is a common language facilitation technique called "expansions."[36] If Stella said "eat," I responded by saying "Stella eat," or "Jake Christina eat." If Stella said, "Come outside," I responded by saying "Christina come outside" or "come outside play." Providing these expansions of Stella's utterances reinforced what she said and also showed her examples of which other words she could add to her phrases in the future. If every time Stella said "outside," I responded with "Stella outside," it would be more difficult for her to learn that she could pair "outside" with other words to say phrases like "outside good," "play outside," "walk outside," "no outside." I varied my responses as much as possible to help Stella understand how she could use words in many different ways.

On the day that I added Stella's six new words, she continued using her original nine buttons consistently before she started exploring the additions. At first, it seemed like she did not care that her board was filled up. Including more words did not change her usage at all. I was happy to see that we could increase her number of buttons without impacting Stella's motor plans of the original words. That is a hallmark of a functional device setup. If Stella had to relearn how to say words every time we increased her vocabulary size, she would be spending all her mental energy on relearning rather than expanding her current communication.

Stella ran to her board and pawed around at her new buttons, hitting three or four in a row without any pause time in between. She cocked her head to the left and right as she listened to the word each button said. She was babbling on her device. All communicators need time to babble and explore. The only way for AAC users to learn how to use their words appropriately is to start out by pushing random buttons on their devices. They need to hear what each word says and observe what happens around them. So if it looks like an AAC user is "just pushing random buttons," that's completely normal and is actually a great indicator that they are on their way to learning how to say those words appropriately.

"Stella Stella Stella Stella Stella," Jake and I heard from the living room.

I peeked around the corner. Stella stared down at her board. She repeatedly lifted her paw up and pushed it back down. "She's learning how to say her name," I said and smiled.

"Good Stella," I modeled. "Love you Stella."

Stella wagged her tail.

Stella explored her own name first before any of the other new words I added. It was a similar reaction to when she discovered she could say "walk" and "beach." She had the power to say the name she heard us call a hundred times per day. "Stella" was likely the single word she heard Jake and I say the most often. It made perfect sense to me that she would be so quick to say it. Toddlers start referring to themselves by their own names around twenty-one to twenty-four months old when they are consistently saying two-word phrases and occasionally using three-word phrases.[37] Stella demonstrated similar patterns at this time.

"Good bye walk Christina," Stella explored with her right paw. She frantically pawed at different buttons, testing out her different words. She walked through the rows to the left side and hit "Jake Christina" with her back paw accidentally. Stella hopped off her board as soon as she heard words she was not expecting. She circled around her device twice, then walked through it again to say, "outside." Stella continued walking to the door, then looked up to me.

"Stella Christina outside," I modeled. "Let's go!"

For the first time in Stella's learning, I could see a clear divide in her skills. Stella knew the left half of the board well. Those nine words had been with her for months, she had relearned their new locations quickly, and she continued using them every day in unique ways. When she marched up to the left side of the board, she pawed each button with such conviction. She often even looked up to me as she was pawing a button because she did not need to visually scan her board anymore. She looked as automatic as a fluent typist writing a paper.

But the right half of the board was new to Stella. Sometimes her paw slipped off a button as she tried to activate it.

She pushed several buttons in a row, searching for the one she wanted. Her back paws kept activating words unintentionally. I knew exactly what was happening. Stella was at different stages of motor learning for different words on her device.

There are three stages to motor learning. The first is called the *cognitive stage* "marked by highly variable performance. The learner may or may not know what he is doing wrong or how to correct his performance and will need assistance." The second stage is called the *associative stage* where "the learner works on refining his skill. He is more accurate with his responses but occasionally makes errors." The last stage is the *autonomous stage* where "conscious thought is no longer needed. The learner is able to perform the action without assistance and often times while being able to divide attention between tasks."[38] Stella was in the autonomous stage for all her original words. But with her six new buttons, she was still in the first stage of motor learning.

Even though I knew Stella understood all the words I added, she still had to learn how to use them herself. Imagine how it would feel to only have access to the left side of the keyboard when you were learning to type. Then, after you mastered it, the right side of the keyboard became accessible. Even though you know the entire alphabet and how to use each letter, you still have to learn the motor plan for how and when to use the right side of the keyboard. It takes time and practice to move through each stage of motor learning.

While Stella was still in the middle of learning how to use her new words, my sister Sarah and her husband, Stephen, visited us in San Diego over a long weekend. I could not wait to show them the board we had set up for Stella, whom they had not seen since before we moved.

Stella thrived with visitors, soaking up all the attention she could. She wiggled her body like crazy, rolled over for belly rubs at their feet, and smiled constantly when Sarah and Stephen first came over. When they sat on our couch, Stella hopped up in between them so they could take turns petting her and scratching behind her ears.

Suddenly, Stella jumped down from the couch. She walked across the living room and stopped in front of her board.

"Love you," Stella said. She walked back to the couch, staring at Sarah and Stephen, wagging her tail. Stella hopped right back up on the couch in between them.

"Oh my gosh, I think she was talking to us," Sarah said. "We love you, too, Stella."

Stella's ears went straight back against her head. She attempted licking each of their faces. She stayed right there, between Sarah and Stephen, looking happy as could be.

"That's the first time she's ever said 'love you' to someone besides us," I said. "You should feel honored."

Stella was perfectly content with all the attention Sarah and Stephen were giving her. She did not need to say anything to receive what she wanted, but she left the couch anyway to walk across the room to let Sarah and Stephen know how much she cared for them. She literally went out of her way to tell them "love you."

She was also clear in her communication. She did not have a word for anyone besides Jake and me, but she used a word in combination with her eye contact and again joining them to specify she was addressing them. Stella did not limit herself to the words she had available. She combined words with gestures to convey even more specific meanings.

Parents often ask me how I can tell if their toddler is saying a word, or just babbling. It can be tricky to know if sounds are random, or meaningful. According to Dr. Erika Hoff, the director of the Language Development Lab at Florida Atlantic University, a word is "a sound sequence that symbolizes meaning and can stand alone."[39] This means sounds or gestures become official words when the child attaches consistent meanings to them. In determining if a sound or gesture can be considered a word, what's happening around the child is equally as important as what the child said. For example, if a baby said "ba" constantly throughout the day, or randomly, it would not be considered a word. I would say he is babbling or exploring sounds. But if the baby said "ba" every time he played with a ball or pointed to a ball, *ba* would be considered his word for "ball" because there is meaning attached to it. The same reasoning applies to AAC users. When AAC users start saying a word relevant to the context, in multiple scenarios, they are demonstrating their proficiency.

About a week after we filled Stella's board, she started using her new vocabulary in appropriate situations. One weekday morning, when Jake woke up earlier than usual, he took Stella on a much longer walk than they normally took before work. When they returned, Stella walked inside, panting and smiling. As soon as Jake unclipped her leash, she walked over to her board. Instead of trying out different words, she stared down at her array of buttons for a moment, like she was figuring out how to say her thought. Stella stepped forward onto the bottom row of her board with her left paw. She lifted her right paw.

"Good Jake," she said.

"Aww," I said.

"Good Jake walk?" he said. "Love you, Stella."

Stella plopped on the floor, continuing to smile and pant.

Later that morning, I left for work before Jake did. He texted me about five minutes after I had gone. She said "Christina bye" and stared out the window!!!

That was twice in a row that Stella had used our names in such specific ways. I was beginning to learn the answers to some of my questions. Stella did indeed use our names, and she did think about us even when we were not home with her. I was eager to come home at the end of the day and see if Stella would say any more of her new words.

Even when we were out of the apartment, I noticed how intelligent and aware Stella was of her environment. That afternoon, I took Stella to Fiesta Island, a massive dog park with miles of beach and large grassy areas. Stella and I were about a fifteen-minute walk from our car when I felt a couple of raindrops. I looked up to see a dark gray cloud passing right over us. I had not brought a jacket or an umbrella, and there were no trees in sight. I picked up the pace. About a minute later, it completely downpoured. The ground became slick with mud, so I could not run without being at serious risk for wiping out. Stella, meanwhile, was sprinting ahead, leaving me behind. Where the heck was she going? She never ran away this quickly unless she was chasing something.

Stella ran up to a woman about thirty feet away who was walking under an umbrella. Stella carried on walking right under it, keeping dry. The woman stopped and turned around to see where Stella had come from. She waved and waited for me to catch up to them.

"I can't believe she ran all the way over here to stand under an umbrella . . . that's one smart dog you have," she said, laughing.

A few nights later, Stella walked over to her board. "Outside," she said immediately.

I stayed silent. I wondered if she would add any more words to her thought. Stella walked all the way around the top of her board to the opposite corner.

"Stella Stella Stella Stella," she said.

I could tell she was not done. She kept her head down and walked to the bottom of her board.

"Walk," she added.

Stella pawed at the door. "Come outside," she said.

Stella looked up to me and whined, which usually indicated that she had completed her message.

"Okay, come on, girl, let's go outside for a walk," I said.

In total Stella said, "Outside Stella Stella Stella Stella walk come outside." She combined nearly all the words she had available to her that related to going outside for a walk. She said four different words, and eight total words. She even used her gesture of pawing at the door in the middle of her phrase to add to her message, which confirmed that the words she was saying were intentional. This was an entirely new level. She was not only saying one or two words at a time now. She flawlessly used every single form of communication and applicable word available to tell me what she wanted. This happened all the time at work with AAC users. It was so common to see kids say something like "all done no stop off finished bye" when they wanted to be done with an activity. They selected every single word that could convey their meaning to make sure they were understood. I wondered if this happened frequently because many of these children were so used to being misunderstood. Stella said her message loud and clear.

The next morning, Stella hopped off the bed and trotted to the living room.

"Outside," she said.

Jake and I were still tucked in bed, not quite ready to get up yet.

"Stella come," Jake said.

It was quiet for about ten seconds.

"Stella bye," she said.

Jake and I laughed.

"Christina can you take her?" he asked.

It was quiet again.

"Jake," Stella said.

I burst into laughter. "She called you out! You have to go take her outside." Jake rolled his eyes and laughed. He found Stella by the door waiting for him.

Stella had officially progressed to a level far beyond what I ever thought was possible. We had a real conversation, across the apartment, with our one-year-old dog. In the past two days, Stella's extraordinary communication events reawakened the desire I had to share my work. I could not believe that something this revolutionary was happening in my home, yet nobody else knew about it. I wanted to find the right way to introduce this concept to the world so that others could teach their dogs and catch a glimpse of the intelligent and complex thinkers that our pets actually are.

After work, I revisited what I had been writing about Stella when I was home sick the week of Thanksgiving. I picked up where I left off and added more information about her communication advancements since then. I called my best friend from graduate school, Sarah, who also loves AAC.

"I've decided I want to start a blog about this whole experience," I said. "I have no idea how I'm going to do it, but the world needs to know what Stella is saying. Can you help me think of a good name? Maybe something to do with Stella?"

"This is going to be bigger than Stella someday," Sarah said. "What about something with your last name? Like . . . Hunger for Words?"

I knew it as soon as she said it. That had to be it.

TAKEAWAYS FOR TEACHING *YOUR* DOG

- **Add names and other nouns.** This can help your dog communicate more specific messages.
- **Test out your dog's vocabulary.** Spend time using your dog's buttons to try saying the common words and phrases you use. If you're able to use her buttons to say a variety of common phrases, that indicates a solid vocabulary selection.
- **Model three- and four-word phrases.** When your dog starts combining words, keep modeling the next level up. A good rule of thumb is to add one word to whatever your dog said. This helps expand length of utterance.
- **Use the stages of motor learning to help you.** Know that it's possible for your dog to be in the beginning stages of motor learning with some words, and automatic with others. Keep modeling and providing cues for the words that your dog is still learning how to say independently and automatically.
- **Give wait time.** When your dog has shown that she is capable of combining words, give her time to do so. Instead of reacting right away to a single word, wait five to ten seconds to see if she will add to her message. Communicating with AAC takes time. Give your dog a chance to finish her whole thought.

CHAPTER 15

Hunger for Words

On my flight from San Diego to Indianapolis for my niece's birthday party, I settled into my seat, expecting my book and music to entertain me for the trip. Instead, I talked to a stranger about my work with Stella, in great detail, for the first time. A friendly, young woman with curly hair and a bright smile sat next to me.

"What's something you're really passionate about? What do you spend your spare time on?" she asked. I could tell this was going to be a great flight. I would much rather engage in a deep conversation than keep to myself for the whole trip. Her warmth and enthusiasm clued me in that she would be receptive to what I was about to tell her.

"I love that question! Well . . . I'm so passionate about speech therapy that I figured out a way to teach my dog how to talk," I said.

Her face froze. I am sure that was the last thing she was expecting me to say. "Here, I'll show you some videos." I explained the concept of AAC and my work as a speech therapist while I rummaged through my bag for my phone. I swiped through several videos of Stella combining words to tell me she wanted to play at the beach, take a walk outside, or wanted to eat.

The two of us spent the rest of the flight brainstorming how I could share my information. She read the articles I had started writing and gave me feedback. On the back of her airline ticket, she wrote down a list of companies and organizations she thought would be interested in this concept of talking dogs. I spent hours with a complete stranger, sharing how excited I was about Stella's communication, where it could go in the future, and how different society would be if we knew our dogs could talk to us. And the concept captivated her enough to spend the duration of the flight brainstorming and dreaming with me.

This was my passion. I wanted so badly to share this information with the world and introduce what was possible. I could not stop thinking about it. I wanted to connect with other professionals who could help me take what I was doing to the next level. I felt like I was walking around with this giant secret, knowing a potential in dogs and a power of speech therapy that others were not aware of yet.

Throughout the spring of 2019, I spent my weekends and evenings figuring out how to create a website. The task sounded easy. People made websites all the time. Surely, if they could do it, I could do it. But I quickly found out it was not easy. Besides creating the actual content that would appear on the website, it required decision after decision about layout, color schemes, font choice, and formatting. And it required competency in a website builder platform, search engine optimization, and a whole host of concepts that sounded like a foreign language to me. I had absolutely no background in any of these areas. I knew my website would not be perfect on my first attempt, but I wanted it to be taken seriously. To me, this was more than a fun side project now. It was my opportunity to introduce a new idea to the world with my professional reputation behind it.

After several weekends spent at my computer, attempting to put something functional and presentable together, I made my blog, www.hungerforwords.com, live on April 25. It had two blog posts, "Teaching My Dog to Talk," which gave an overview of the process from the start of my idea through Stella combining words, and "Stella's Buttons" about the words I chose to teach Stella, several videos dating back to her first words, information about me, and information about Stella. It was simple, colorful, functional, and the first time I had ever created something like this on my own. About one year ago, Stella said her first word. And now, here I was, living in a totally different part of the country, launching a website about her communication skills. So much can change in twelve months' time. I was excited to have a space to share information and inspire myself to keep pursuing this journey. Even though I knew nobody was aware of my blog yet, it still helped me feel a sense of legitimacy and responsibility. This was my vision. I was responsible for bringing it to light and sharing my observations with the world.

First, I only shared my blog with people I knew. I emailed it to my family, friends, professors and supervisors from graduate school, current and former coworkers. Seeing their enthusiastic reactions helped me work up the courage to share it with a slightly wider net of people, the speech-language pathology Facebook community. I had not been on social media in months. Ever since I deleted the apps from my phone before moving to California, I never looked back. I felt so free without social media overtaking my spare moments and time to think. I was hesitant to log in again. I did not want to slip into old habits. But my desire to share Stella's story and the power of AAC was far greater than my fear of becoming addicted to social media again.

A week after the website launch, I posted the link in a couple of SLP Facebook groups with thousands of members in each of them. I had never posted in one of these groups before. I figured this would be a great starting point to gauge reactions. If anyone should understand and appreciate what I was doing, it would probably be the speech therapy community. It was a safe way for me to test the waters and see how others would perceive my work.

"Dogs can use AAC too!" I wrote. "Check out my website to see how my dog, Stella, is saying words." A picture of Stella lying down next to her buttons, smiling, accompanied the caption.

I sat across our dining table with Jake when I clicked "post." I didn't know what to expect. "I hope at least a few people see it," I said. "There are so many posts in these groups. Things can become buried."

"I don't think you have anything to worry about," Jake said. "Look." He turned my laptop screen toward me. In a matter of minutes, hundreds of "likes" and comments swarmed my post.

Over the course of the night, other speech therapists started sharing my link on their own personal pages. A couple of large speech therapy accounts with tens of thousands of followers shared it to their social media platforms. People were signing up to receive blog updates. A community of people interested in my project was forming. The interest was officially there and spreading.

Stella started to learn when to ask for "beach." She used to ask to go constantly, morning, afternoon, and night. But we could never go to the beach in the mornings and make it back in time to go to work. It was the same with my lunch breaks. I could never say yes to taking her if she asked when I stopped home in between clients. The only times we ever could honor her request were after work and after dinner. Stella started adjusting her communication when she realized these patterns. In the mornings and afternoons, she stuck with saying "outside" or "walk." At the end of the day, she would request the beach. As she learned our routines, she adapted.

Every week it seemed like Stella was surpassing the skills she demonstrated the week prior. Three-word combinations became frequent instead of occasional. Stella was averaging approximately thirty different utterances per day, even when we were only home for a couple of hours in the morning and a few hours at night. Something undeniably special was happening and the scientist in me took over. It was time to track and document her progress more thoroughly. Jake and I created charts that we hung on the walls to mark every time she said a word. We recorded stories of her communication, including the context of what was happening at the time, so it would be meaningful to us later on. I jotted down questions I had about her progress and tracked days of her language samples to look for patterns and assess her skills. We took videos with our phones and experimented with having a GoPro camera set up in our living room to catch everything Stella said. We spent hours combing through video footage, splitting clips, and storing them on our computers. I continued writing blog posts about Stella's communication and practicing words with her. In the middle of my speech therapy sessions with toddlers during the day, epiphanies popped into my head about what Stella had possibly been trying to communicate earlier. When I was at home working with Stella in the evenings, I remembered situations from the workday that led me to try new concepts with her. Could she answer simple questions? Could she make choices if I gave her two options? What if I asked, "Stella want play or Stella want walk?" Would she answer with "play" or "walk"? Could she understand the difference in my intonation when asking a question in comparison to telling her what we were doing? There was still so much to discover.

It was a busy time, so Jake and I decided to take a vacation. Since she was too big to fly, Stella stayed with a dog sitter in San Diego for the week. When we met the young woman a few days before dropping Stella off, I listed what we would make sure to bring.

"We'll have her kennel and bed, and her favorite blanket, oh and . . . Stella also has a communication device with buttons on it that she uses to say what she wants," I said.

She frowned. "Okay, yeah, there's plenty of room here for all that."

A few days later, I showed up on her doorstep again, this time holding Stella's leash in one hand and her board of buttons in the other.

"So this is her device," I said. "Stella will tell you what she wants and what she's thinking." I pushed down on a few buttons to show the sitter how it worked.

"Oh . . . wow, okay . . ." she said, scratching her head.

"It should be really convenient. She'll let you know if she's hungry or if she needs to go outside, or where she wants to go." The sitter nodded along while I rambled on about all the things Stella tells us every day.

"Bye, Stella," I said. "Love you, girl. Jake and I will come back soon. Have fun!" I kissed her head and headed out the door. It was so hard to leave her. I hoped she would be happy and feel comfortable while we were gone.

Ten minutes after I left, my phone lit up with a text from the dog sitter: Okay, I'm really spooked . . . Stella just said "Christina bye" . . . I laughed to myself. I wish I could have been there to see the sitter's face in person when Stella marched over to say that I had left. I was glad to hear that Stella was already using her words.

While we were away, we received a video of Stella playing on the dog beach. She kept saying beach, so I figured we had to take her, the sitter texted. We also learned that on the first night of her stay, Stella said "outside" in the middle of the night. The dog sitter heard her from her bed but did not get up. Stella said "outside" a couple more times before she went to the bathroom next to the door. As soon as that happened, I realized she really meant what she was saying, the sitter wrote. That accident was completely my fault.

On the last day of her stay, the sitter packed up Stella's bed, toys, and dishes. Stella watched her compile her belongings, then said "Jake Christina." Apparently, when the sitter told her, "Yes, Jake and Christina are coming back!" Stella waited by the door until we arrived ten minutes later. I wonder if Stella understood her words or had learned that pattern from staying with our friends in the past. They usually packed up her things right before the two of us came back to pick her up. Regardless, Stella's words helped her navigate the situation of staying somewhere new without us. She let the dog sitter know when she realized I was gone. She was still able to communicate all her typical wants and needs. And, with the buttons sticking to her board, she did not have to get used to a new layout or figure out which word was which in a new location. She could carry on, communicating as usual to new people in new places. This all mattered, especially since it had become really hard for Jake and me to leave Stella. She was more like a child than a pet to us. She had her own voice, and thoughts, and had become such an important part of our lives. It comforted me to know that Stella had the ability to voice her thoughts when we were not there.

Back home, I decided to upgrade further, to a larger board. Stella had learned the words on her current device and used all fifteen of them appropriately and frequently. As I've said, when possible, AAC users should always have access to more words than they know how to use. The only way to learn new words is to have access to saying them.

Jake and I took Stella to Home Depot where we picked out a piece of plywood a little over twice the size of her current poster board. Jake measured everything out. "It would fit thirty-two total buttons. I don't know if we'll get to that point but at least we could," he said. Stella trotted next to us, galloping toward any person she saw with a friendly face.

When we returned home, we arranged Stella's fifteen buttons on the larger board, starting again from the top left corner and keeping all her words in the same order. This board was longer and wider. It could fit a fourth row of words on the bottom. I consulted the list I made the last time I expanded Stella's device and selected the six that I thought would be the most impactful and meaningful for her: *want*, *look*, *park*, *happy*, *mad*, *bed*.

Jake and I frequently asked Stella "What do you want?" So I know she heard the word *want* often. I unintentionally verbally modeled "want" all the time whenever Stella made a request. When Stella said, "eat," I responded, "Stella want eat?" When she said "beach," I said, "You want beach?" Now I could model these phrases while using her buttons for each word. *Want* is an excellent core word that can apply to many different objects, actions, and people. If Stella learned how to say the word *want*, she could also combine that with a gesture to tell me what she wanted instead of taking up room on her device for several specific toys or activities.

One of Stella's favorite indoor activities was looking out the windows. Whenever I opened the shades for her, I narrated what Stella was doing by saying, "Stella look outside." I wanted to give Stella the chance to talk about looking outside or let us know that she wanted the blinds open. I also said "Stella look" anytime I pointed something out to her. Stella would follow my point and look to whatever I was trying to show her. Research shows that puppies as young as six weeks old respond to human social gestures such as pointing.[40] Maybe Stella would use "look" to point things out to me, too.

Stella and I started adding trips to the local dog park into our mix of evening activities. If I said the word *park*, she ran to the door. She loved running around with her dog friends and teasing them to chase the stick dangling out of her mouth. Since we usually went at the same time of day, we ran into the same people and dogs all the time. Stella greeted everyone happily, and sometimes I wondered if she wanted to see her human friends more than her dog friends.

Stella demonstrated plenty of gestures that indicated if she was happy or upset. When she was happy, she wagged her tail, her ears went back against her head, she jumped around in circles, and she smiled. When she was upset, she barked at us, sighed, turned away, or walked into the other room. Since she was expressing such clear gestures about how she was feeling, I wanted to give Stella the opportunity to tell us more about her emotions.

Every day, we pulled Stella's bed to a variety of locations in the apartment. She liked when we set it under the coffee table for her, by the window, or at the foot of our bed. She often stood where she wanted her bed and whined or barked at us. Once again, since she was already vocalizing and gesturing about this concept, I figured she would have success with using the word *bed* to talk about what she wanted.

I lined up the six new words on the bottom row of Stella's new board. Jake and I added labels under each word like we had on her smaller board. The labels helped visitors and people like the dog sitter identify buttons if they wanted to talk to Stella using them. We decided we would wait to Velcro the buttons down to the board until we knew for sure we were keeping this size. I kept the original poster board between the wall and the desk in case Stella did not do well with the bigger board.

"Look, Stella," I said, modeling with her buttons. "Come look."

Stella walked back and forth through her new board a couple

of times. She sniffed it, looked at me, then walked over to my desk. She started pawing at her old board.

"Want," I modeled.

I pulled her old board out and laid it down on the ground. Would she paw at the spaces where the buttons used to be, to show me which word she wanted to say? Or was she trying to tell me to put her buttons back on this board? Neither, actually. She lay down, sprawled across her old board. She stayed there for the rest of the afternoon. That board was Stella's. She was physically hanging on to it, stopping us from putting it away or throwing it out. She seemed reluctant to let it go, like she needed transition time before she could start making the new board hers.

Once again, I thought back to children switching communication devices. It must be so hard for them to let go of a device that had been their voice for years and try learning a new system. We all need support in times of transition, especially when dealing with something so personal as one's method of communication.

When Stella stood up from her old board, I modeled all Stella's words for her, showing her that all the same words were still there, in their same locations. I modeled the new row of buttons for her as she watched and listened to each word. The transition to this new board looked like it was off to a much better start than before. She did not retreat to the couch or bed like she had the first time we placed all her buttons on a board. She watched me and actually stepped right up to test it out.

"Want," Stella said. She continued walking up the board. "Outside."

"Stella want outside? Okay, let's go outside."

Stella picked up on "want" incredibly quickly. From this very first attempt using it, she continued pairing "want" with her requests to go outside, play, eat, go to the beach. Since Stella had heard us say "want" so frequently, and since Stella already had such strong abilities to combine words, she could incorporate "want" into her vocabulary seamlessly.

Stella walked back inside smiling. "Stella happy," I modeled. "Happy happy." Stella continued smiling and plopped down on her bed. "Stella bed," I modeled. "Love you, Stella." Stella's tail wagged.

Over the next couple of weeks, Jake and I modeled Stella's new words. Before I took Stella to the dog park, I said "park." Every time Stella lay on her bed or on our bed, I modeled "bed." If Stella requested going to the beach when we could not take her, I modeled "mad" when she whined at us. I modeled "look" every time I pointed at one of Stella's toys, or whenever Stella was looking out the window. Stella continued using her old words and started incorporating her new vocabulary into her phrases more and more each day.

Almost every night, Stella let us know when she was heading to bed. Jake and I would be in the living room talking or reading. Stella would say, "bed," then walk past us into the bedroom to fall asleep. And when Jake went out of town for a few days, Stella said, "Jake no bed," before she hopped up on his side of the bed and slept there for the night. If the door to the bedroom door was closed, she would say "bed" or "help bed," then try pawing it open. The word *bed* became instrumental in my awareness that Stella was trying to communicate a sequence of activities to us. She would wake up in the morning, say "bed eat" or "bed outside." In other words, "I'm done sleeping, now I want to eat." She eventually added three words to the sequence to say phrases like "bed eat outside" or "bed outside eat." Or she would come in from me taking her out first thing in the morning then say, "bed outside eat," letting me know all the steps of her morning routine, spontaneously.

Every time I add new words to Stella's board, she exceeds my expectations in many ways. I may think of a few reasons why it would be a good idea to include a certain word, then picture a few different scenarios where I could see it being helpful. But Stella uses words in ways I would not have thought of or combines them to create such unique phrases. This is further proof to me to not direct Stella what to say when. When I focus on modeling words in a variety of contexts, she learns the meanings of them and decides how to use them for herself. She does not need me telling her exactly what to say when. If anything, that stunts her growth more than it helps it.

Stella commented on unique situations that happened in our apartment that I never could have anticipated when I first imagined how she would use words to communicate. One day, our neighbors were dog sitting. When Stella saw them walk past the window with an unfamiliar dog, she barked, said "help no help" and ran to the window to bark again. Stella recognized that our neighbors were bringing a stranger into our complex. She did not bark when they walked with their own dog into their unit. I wondered if Stella was using her words to tell us why she was barking.

Stella's social use of words continued to grow with her vocabulary skills. She started repeating herself if we did not respond to her. One morning, she said, "come come outside." When Jake and I continued our conversation and did not stop to respond to Stella, she repeated "come come outside" and stared at us. I wondered if she learned to do this by observing us repeat a question to Stella if she did not respond. This is such a strong social skill. Stella had the awareness that she said something, we did not respond to it, so maybe she should try again.

Jake and I sat at our table, eating dinner after work. Jake was in the middle of telling me a story when we heard, "Beach eat come eat come," from the living room.

"Yes, we'll go to the beach after we eat, Stella. Eat now, beach later," I said.

"Come come come come," Stella said.

"Just wait a little bit longer, Stella."

Stella stayed by the door, monitoring our meal progress from afar. As soon as we took our dishes over to the sink, she wagged her tail and waited by the door for us. When we came back from the beach, she walked inside and said, "Bye Stella bye good outside." She was using her words to comment on what she just did, not what was happening now or what she wanted.

The next day, I took Stella to the dog park after work. She played with her favorite friends, two Rottweilers who always engaged in tug-of-war with a stick for as long as Stella liked. By the time we came home from the park, Jake was already back from work. Stella ran inside, greeted him, then said, "park play." *Is she trying to tell Jake that was where we were?* I wondered. *Is she trying to share a story about playing at the park, but she doesn't have any more vocabulary to use? Is she saying she wants to go back to the park?* This situation was coming up more and more. I knew we needed a way for Stella to distinguish between something she just did and something that she wanted to do.

TAKEAWAYS FOR TEACHING *YOUR* DOG

- **Give your dog transition time to adjust to a new board.** Even if the words are all staying in the same relative locations, your dog may need time to adjust to a new board. You can make it easier on her by keeping her old board nearby for a little while until your dog claims the new one.
- **Model emotion words when you see your dog exhibiting the emotion.** When your dog is smiling, wagging her tail, jumping in circles, or playing at her favorite place, use these opportunities to model the word *happy*. When you can tell your dog is frustrated or upset, use these times to model the word *mad*.
- **Avoid telling your dog when to say certain words.** Constantly telling your dog "say outside" or "say good" will teach your dog to say what *you* tell them to say, not to use the buttons for what they are thinking. We are teaching our dogs how to use words, not training them to talk on command. Modeling and naturalistic cues are most effective.

CHAPTER 16

Language Explosion

"Stella bye play," Stella said. Jake and I were eating dinner. Stella stood next to her buttons, staring straight at us.

"We're going to eat now, Stella. We'll play later," Jake said.

Stella sighed and whined. "Eat eat park."

"Yes, we're eating now, then we can go to the park," Jake said.

Stella lay down in front of the door.

Stella desperately needed a way to communicate about time concepts. Sharing these little sequences of what we were doing now and what Stella wanted to do next were daily occurrences.

The next day, I added *all done*, *now*, *later*, to Stella's board. I modeled *now* right before or during an activity. I would say, "Play now" as she was playing, "eat now" when she was already eating, "park now" when I put her leash on and grabbed her ball. I used *later* to talk about anything happening later than the next ten minutes. I modeled *all done* when Stella finished any activity. Saying "all done eat" when she was finished eating, "all done water" when she stopped drinking water, "all done play" when she dropped her toy to lie down, and "all done park" when we came home from the park all helped Stella understand that *all done* meant the conclusion of something. Even though I just added these three words, I had always used them in my natural vocabulary when I talked to Stella. Stella was no stranger to these words; this was just the first time she had the chance to use them as well.

The combination of time concepts and two emotion words, *happy* and *mad*, caused another language explosion. After only a few days of modeling *all done*, *now*, and *later* whenever possible, Stella started incorporating these concepts into her own phrases.

One evening, I started vacuuming our apartment, which she always hated. She typically ran into the other room and peeked her head around the corner to watch me carefully from a distance. Or she stood on the couch or bed while looking down at the vacuum to stay out of its way. This time, after I was vacuuming the living room for about three or four minutes, Stella ran from the bedroom over to her board. She sped right past me, barely making eye contact with me and avoiding the vacuum.

"All done all done," she said.

I turned the vacuum off.

Stella wagged her tail. Her ears went straight back against her head. "Happy," she said.

"Aw, you're happy it's all done? Good Stella, good girl." I petted Stella and put the vacuum back in the closet. It could wait until Jake took her on a walk later.

This was the first of many times that Stella used "all done" to tell Jake or me when she wanted *us* to be finished with something. She directed us more than she used "all done" to narrate her own activities. If we lay in bed for longer than usual, she told us "all done," and whined from the living room. When we took longer to eat than normal, she said "all done," then walked into the kitchen. It looked like she was trying to cue us to take our dishes into the kitchen like we always did when we finished our meal. If I was writing reports on my laptop and had ignored Stella's play attempts, Stella would tell me, "all done." Almost every time I was talking on the phone, Stella became her chattiest. She would repeatedly say "all done," along with what she wanted me to be doing with her instead. This happened at work with toddlers every single day. As soon as the parent and I would engage in a longer conversation at the end of the session, the child would do whatever he could to grab our attention. Seeing how often Stella told us "all done" made me wonder how else she might be wanting to direct us.

In the summer of 2019, Stella was using over twenty words independently and functionally. She combined words several times each day and continued to make novel phrases. Stella used words to request actions, request places, narrate her morning routine, call out to Jake or me, talk about what just happened, and share how she was feeling. The patterns of language she was using were consistent and predictable. She said words at the times when they made sense. She did not say "eat" in the middle of the day or combine words that did not go together like "beach bed," "good mad," "bye bed," "water park," "walk water," and so on. She had reached the automatic stage with all her words, and it showed. She communicated in several different environments with different people, not only to Jake and me. She first learned to use her buttons on their own, and now on two different-size boards.

Stella's vocabulary was equally as impressive to me as her social skills were. Stella's social use of the buttons matched many of the same social rules we use while we talk. She tried getting our attention first by saying "come" or "look" if we were in the other room before she carried on with her message. She made eye contact with me after she finished her thought, then waited for a response. If we did not understand her, she tried saying what she meant in a different way, or she repeated her message.

Stella rarely interrupted me. When she was emotionally charged, she pressed buttons with more intensity and more repetitions to stress her point. With verbal speech, we can adjust our tone and volume. AAC users, however, do not have that same privilege. AAC users often convey different tones by the way they press buttons, or with the gestures they use in combination with their words. This is similar in American Sign Language. Emphasis is put on words by the speed and force used to make each sign, and the facial expressions of the speaker. When Stella was tired, she stretched along her board, saying words slowly with significant pause time between each one. It reminded me of when I wake up in the morning and yawn and stretch as I talk. When Stella was frantic, she ran to her board and slammed on each button, just like someone who is so excited that they're blurting the words out. Language is so much more than knowing the meaning of words or knowing which button says which word.

One day, Stella's toy landed on our built-in bookshelves. She stood on her hind legs to grab it, accidentally knocking down a sign in the process. Stella looked to me and became submissive. Her tail tucked down between her legs and she lowered her head. She walked to her board.

"No," she said. Stella looked at the sign she knocked over and back at me again. She walked over to me, slowly wagging her tail.

It was her way of saying she didn't mean to do it. "It's okay, Stella! Everything is okay." After some verbal reassurance and petting, Stella perked up again and continued to play.

Stella listened when Jake and I talked between the two of us. One night, we sat on the couch talking about if we wanted to go outside again or if we were done for the night. "I think we can stay in," Jake said. "She's probably good." Stella hopped off her bed and walked over to her board.

"Mad outside come come outside," she said. When we did not jump up from the couch immediately to go outside, Stella barked, then said, "Walk mad Jake outside." Stella did not want to wait for us to decide what we were doing. She spoke her mind, letting us know her say in the matter.

Stella even came up with her own ways to talk about concepts she did not have a specific word for. For example, she started saying, "bye eat" to request the Kong toy we filled with peanut butter every time we go to work. She did not only say this once or twice. She said it nearly every day when we were preparing to leave, or if it took us longer to leave than usual.

"Bye eat?" I asked. "You want to leave and eat somewhere else? Or we are going to leave then you will eat when we come back?" I stared at Stella, trying to figure out what she meant.

Stella licked her lips. I grabbed my bag, lunch box, and water bottle, tossed Stella's Kong into her kennel for her, and rushed out the door. Ten minutes into my drive, the answer popped into my head. *Duh*, I thought. *She was trying to tell me she wanted me to leave so she could have her peanut butter.*

This was yet another important lesson to not dismiss communication if we do not understand it right away. Just because the adult may not connect the dots immediately, it does not mean the communication was random or unmeaningful. This happens all the time with toddlers. The last time I was with my one-year-old niece, Clara, I watched this exact same scenario unfold. Clara sat in her high chair, waiting for her food.

"Paper towel," she said.

"Here you go." I set a paper towel down on her tray.

"Oh, she wants a clementine," Kate, my older sister, said.

"What?" I asked.

"Every time Clara eats a clementine, she likes putting the peel on a paper towel."

Kate was very aware of Clara's routines, which allowed her to understand this communication without skipping a beat. Clara knew how to say "paper towel" but did not know how to say "clementine" yet. She said what she could to communicate her desire. If I had been alone with Clara, there was no way I would have jumped to the understanding that she wanted a clementine. Just because I did not understand did not mean it was unmeaningful. Sure enough, when Kate opened the fridge and pulled out a clementine, Clara started clapping.

"Beach play," Stella said. I was in the middle of making dinner. Stella kept asking for the beach, but I knew the three of us were going to visit the beach later that night after dinner.

"Beach play later, Stella," I said.

Stella barked at me. I walked over to her board.

"You sound mad, girl." I said. "Stella mad. Beach later," I modeled.

Stella stood next to the door whining. I returned to the kitchen, but I could still hear her in the living room. I stirred the soup on the stove and thought back to my speech therapy sessions with toddlers. If a child asked for something we could not play with at the time or asked to go somewhere, I would not only say "no" or "later" and leave it at that. I would have offered the child something else to do instead. Giving the child options for activities we could do in the moment helped him feel in control and showed him that there were other fun toys we could use. Maybe Stella needed a little more direction. Maybe she would benefit from having options too.

I walked back into the living room. "Beach later Stella," I said. "Want play toy now?" I modeled "want play now" while pointing to Stella's toy bin. "Or want bed?" I modeled "want bed" and pulled her bed out to the kitchen so she could lie and watch me cook if she wanted. Stella became quiet. She looked around the room and back at me. She picked up a toy, squeaked it, then dropped it. Stella walked over to her bed. She plopped right down and watched me finish making dinner. "Good Stella," I said. "Good Stella on bed now."

With so many frequent communication advancements, I created Hunger for Words Facebook and Instagram accounts. I had continued writing blog posts and updating a couple of new featured videos on the homepage of my website every couple of weeks. My audience was mostly friends, family members, and friends of friends. But every so often, I would notice some strangers stumbling upon my blog and following along to see Stella updates.

I was hesitant to return to the world of social media. I had created the space I needed in my life to be able to progress this far with Stella and focus on what was important to me without mental clutter standing in the way. But I also had so many videos of Stella talking that I wanted others to see. It was fine for people to read about her communication stories through my blog posts, but it was always more striking actually seeing her in action. I hoped these videos would help people start thinking about communication potential in both humans and animals in a new way. I started sharing videos of Stella talking, and writing captions to explain the skills she was demonstrating and what I was noticing.

In July, only four months after launching my website, and six months after I moved all Stella's buttons to her first board, I received an email from a writer at the alumni magazine at my graduate school, Northern Illinois University. She had heard about my work with Stella and asked to write a piece on me.

"My first interview!" I shouted to Jake.

"The first of many, I'm sure," he said.

I could not wait for the chance to talk about speech therapy, AAC, and Stella's success in a way that could potentially reach a large audience. The writer sent me the list of questions she was planning to ask. I mulled them over for multiple days, practiced responses, and wrote down talking points. I wanted to represent my work in the best way I could.

After the piece came out, I gained more followers on my social media pages and email subscribers. I received a few thoughtful emails from fellow alumni, sending well wishes and excitement about my work. Now a few hundred people were following us along on our journey. I wanted this story to spread because of the messages it represented. Everyone deserves a voice. And dogs say and think about so much more than we ever thought they could.

At the end of the summer in 2019, we moved to a new apartment near the dog beach. It was heaven for Stella. There was a giant park across the street from us on the right, and the path to the dog beach on the left. We drove to this area all the time after work and on the weekends anyway, so we figured we might as well live where we spent the majority of our free time. I wondered how Stella would react to moving again. I knew it would not be nearly as traumatic as moving across the country, but I was worried she would panic seeing our home emptied out. I wondered if she would use her words at all during this transition. Last year when we moved, she needed some adjustment time to become comfortable in her new space before she said much. But now, Stella was older, had significantly stronger communication skills, and a board that kept her words in the same locations. I was eager to see how she would do.

On the first evening in our new apartment, we emptied out half our boxes. Stella sniffed everything we unpacked and carefully watched us set up our new space. Her gaze tracked our every movements. This apartment was still a one-bedroom unit, but it was slightly larger than our last one. The twelve-foot-tall ceilings and large windows in the living room made it feel much more spacious. We placed Stella's board a few feet away from the door, at the edge of our living room.

"I want to show her where we can walk to now," I said. "Let's take her to the beach!" Stella came running over to me. She must have heard her favorite word. "Yes, beach. Let's go to the beach, Stella."

As soon as we reached the elevated bike path that led to the dog beach, Stella smiled and bolted for the shore. She and I ran together, both equally excited about our new neighborhood. When we returned from our sunset playtime, Stella walked straight to her board.

"Beach play love you," she said.

"She is going to be so happy living here," Jake said.

The next Saturday morning, Jake and I woke up early to resume unpacking and rearranging our furniture. We were debating on the location of our couch and had moved it back and forth across the living room three or four times by now. Each time we lifted the couch, Stella jumped back as far away from us as she could go while still keeping an eye on us.

"You're right, I think it's better over there," I said. "Let's move it back. Final decision."

Jake and I took our positions. We squatted down to lift it. Jake started counting. "1, 2, . . ."

Stella barked and ran to her board. We paused.

"All done walk happy walk happy want," she said.

"I know you want us to be all done now. Hang on, Stella, walk later," I said. We moved the couch across the room and continued unpacking for a few more minutes. Stella walked back to her board.

"Stella bye love you," she said. She sat in front of the door, leaning against it. She was absolutely ready to leave the chaos.

The location of our new apartment brought even more understanding of the meaning behind Stella's words. Across the street, there was a clear fork in the road where the beach was to the left, and the park was to the right. If Stella said, "park," but I really wanted to go to the beach and tried walking her that way, she would lie down in the middle of the sidewalk until I turned the right way. Even though she did not have her board out on walks with her, she found a clear way to tell me that she meant what she said. I learned that the beach and the park were not equal experiences for her. I do not know exactly what made her more in the mood for the park instead of the beach some days, and vice versa, but she clearly had a reason and a preference, like we all do. Maybe some days she wanted to see the squirrels at the park, and on others she wanted to chase birds down the shore. Maybe some days she felt like rolling around in the grass, but on others she wanted to dig in the sand. I wondered if she had more vocabulary available if she would be able to tell us exactly what she wanted to do during her playtime.

Stella's commitment to her own wants and needs inspired me. She did not back down from her own desires when I tried taking her where I wanted to go. This was further proof to me that Stella's sole purpose in life was not to do what we say. She had her own mind and ideas. Everyone has the right to their own opinion and their own wants. Now that I saw first-hand that Stella had a specific vision for her playtime, and did not appreciate me trying to change it, I tried finding a balance between what Stella wanted and what Jake or I wanted and talking about it first. Before we left, I would model "Christina want beach," or "Christina want walk." Then I would ask, "What Stella want?" We always took Stella's wants into consideration like we would with another human. Some days we all had the same wants, other days we picked where Stella wanted to go, and sometimes we chose where Jake or I wanted. When we told Stella beforehand where we were going, she would not lie down in protest. She just needed us to tell her what was happening so she could understand.

I opened another new box of Recordable Answer Buzzers. Now that Stella used all her words consistently, it was time to add more buttons for her. Stella's behavior around her board always showed me when she really needed an expanded vocabulary. After she had learned all her words and used them all independently, sometimes she walked through the rows and whined or stood on her board whining. This made me wonder if she was trying to tell me something that she did not have a word for. Stella watched me remove the buttons. Her tail wagged and she licked the box.

"Happy want," Stella said. She walked back to me and licked the box of buttons again.

This time, I added fringe words for Stella: *ball*, *toy*, *couch*, *inside*. Since Stella was communicating with such ease and at such high frequencies, I figured it was time for some more specific vocabulary to help her make her messages even clearer. I modeled phrases like "Stella play ball," "play ball inside," "want ball," every time we played with a ball. I said, "Stella play toy," "play toy inside," or "help toy couch" when her toy slid under the couch.

One weeknight at the beach, Stella found a ratty scrap of some other dog's toy buried in the sand. It was disgusting. But Stella would not let it go. She carried it the whole way home. When we reached the sidewalk outside of our apartment, we told Stella to "leave it," and ran inside before she could pick it up again. When we walked inside, Stella stood by the front door with her nose pressed against it. She kept sniffing and pawing at the door. A few minutes later, Stella walked over to her board.

"Toy inside," she said.

"You want to bring the toy inside?"

Stella wagged her tail and ran back to the door.

"Let's play toy outside, Stella. Come on, girl." I clipped Stella's leash on again and took her back outside. Stella pounced on the toy. She clenched it in her mouth and shook it back and forth. I still did not want to bring it inside, so I let her play for a few more minutes outside before being all done for the night.

While Stella's vocabulary was growing, the situations she talked about were also becoming more and more complex. New and unusual circumstances led to greater revelations about Stella's cognition. One morning, Stella accidentally bumped the reset tab on her "beach" button. It made a loud sound, then erased the word.

"Mad," Stella said.

"I know, Stella, let me fix it. Christina help." Stella tracked my every move. She always watched closely when I intervened with her device. I picked the button off her board and tried recording "beach" again. But it didn't work. The button had broken after so much use.

"Sorry, girl," I said. We didn't have any spare buttons.

About five minutes later, Stella approached her board. She put her paw on the empty space where "beach" was supposed to be, then sniffed it. Stella paused on her device, looking around to her other buttons.

"Help water outside," she said.

"Oh my gosh . . ." I said.

Stella had the understanding and the problem-solving skills to figure out a way to talk about her broken "beach" button. This was the first time we had a malfunction with her "beach" buzzer. We had never practiced anything like this or talked about this situation. Stella was using words in unique ways during brand-new circumstances.

A couple of days later, we ran out of Stella's typical food. I took it upon myself to make her a special dinner instead of running out to buy more food right away. I plopped the rice, chicken, and vegetable medley into her bowl. "Here you go, Stella."

Stella sniffed her food then took a couple of small bites. She looked up to me.

"Stella eat," I said.

Stella sniffed her food again, then walked to her board. "Eat no," she said. Stella hopped on the couch and curled into a ball. I assumed she would have enjoyed a different (and home-cooked) meal. But she didn't go near it the rest of the night.

The next day, after we bought more of her normal food, Stella scarfed down her bowl then walked to her board.

"Happy eat," she said.

I was glad Stella had a way to express even simple opinions about what we fed her every day. I really didn't know she enjoyed her specific food so much until she said "happy eat."

A couple of weeks later, Jake, Stella, and I spent a weekend visiting several of Jake's friends. When we arrived to a living room full of people, Stella ran to greet each person. I set her board down against the wall while Stella enjoyed all the attention. Everyone formed a circle around her, taking turns petting her and giving Stella lots of love.

I stepped into the circle. "Stella, come here," I called. I wanted to pet her too.

Stella glanced over to me, then walked to her board.

"Christina later," she said.

My jaw dropped. Jake burst out laughing. Our dog just told me she wanted to hang out with her new friends instead of me right now.

It felt like we hit a sweet spot with the now twenty-nine words on Stella's device. We had a great combination of core words and fringe words for her to be able to talk about most of her daily activities. The number of words is not everything. Having a solid group of vocabulary options available for Stella to talk about several different situations is far more important than having tons of words that do not really impact how she communicates. But remember, we all know thousands of words, yet the same three hundred to four hundred words make up 80 percent of our everyday speech.[41] I still wanted to add more words to the empty space on Stella's board, but I was so happy with the balance of words we had achieved so far.

I experimented with some words along the way that I did not end up keeping on her device because they did not serve a unique purpose for Stella. I tried *get*, *go*, *kennel*, *then*. I thought *get* might be a useful addition when I realized how often I say "get your toy," or "get your ball." But Stella ended up using *want* in all the scenarios where *get* would also be appropriate. *Go* is a great word, which I would normally always recommend adding, but Stella had learned to say the specific places she wanted to go to already. Also, she had started using *bye* like *go*. If I were starting over again, I probably would have incorporated *go* earlier on. Stella used *bed* to talk about her kennel since her bed was in there. And *later* was more functional for us than *then*. We could use *later* on its own, like "beach later," or in combination with what we were doing now—"Eat now beach later." But we could not use *then* on its own, only in combination with what was happening now. It would not make sense to just say "then beach."

Little differences in vocabulary uses like these can go a long way in picking which words are on a communication device. In a perfect world, we could have all the words available for Stella as we do on human communication devices, but we are not at that point yet. So for now, selecting the words that gave Stella the most bang for her buck worked out well.

TAKEAWAYS FOR TEACHING *YOUR* DOG

- **Introduce time concepts to help your dog differentiate between something that already happened, is happening now, or will occur later.** Model "now" in combination with what you are currently doing and "later" if you plan on doing a specific activity later in the day. If your dog says "now," try as best you can to react quickly!
- **Give your dog a way to say "all done."** Having words like "finished" or "stop" or a phrase such as "all done" gives your dog the power to say when she wants something to be done. Model "all done" at the conclusion of your dog's activities to help her learn its meaning.
- **Talk to your dog about what's happening.** I learned with Stella that if I told her what we were doing beforehand, she was less startled, and reacted more calmly to changes in her routine, or an activity that was different from what she requested. We are all more understanding when we know what's going on.

CHAPTER 17

Hello, World

At sunrise on October 30, 2019, I opened my journal and wrote on the top of the page, "Wouldn't it be amazing if . . . ?" This is one of my favorite early-morning activities, letting my imagination run wild before the rest of the world wakes up. "Wouldn't it be amazing if my Stella posts attracted hundreds and thousands of new fans bursting with enthusiasm? Wouldn't it be amazing if someday I got to speak to large groups of people about my work with Stella? Wouldn't it be amazing if someday I was doing interviews on TV about Stella and AAC?" I sat with those ideas, completely crazy to me at the time, and let myself feel how wonderful it would be to experience all that.

The next day, I parked my car in front of a beachside apartment building with a few minutes to kill before my next speech therapy session started. I checked my email to see one new message with the subject line "People Magazine Calling!" I opened the email, expecting it to be some sort of spam. My eyes widened when I saw that it was not spam at all. It was a legitimate email from a *People* magazine writer, Hilary Shenfeld, who happened to see my videos of Stella on social media. She wanted to write about Stella and me.

I freaked out. Barely anyone knew about my work with Stella. A lot of my friends did not even know what I was doing. But suddenly, one of the few hundred people in my social media audience happened to be a writer for *People*? What are the chances? I took a screenshot of the email and sent it to Jake, along with a row of exclamation points.

Three days later, on the eve of the scheduled interview, I opened a text from one of my SLP friends: You didn't tell me you were going to be in the Leader this month!! it read. The *ASHA Leader* is the professional magazine for speech-language pathologists and audiologists.

What?? I didn't know I was going to be! I haven't gotten my copy yet, what is it? I texted back.

She sent a picture of one of the first pages of the magazine, a column written by Dr. Shari Robertson, the president of the American Speech and Hearing Association at the time. I loved her work. I ripped out her previous columns and saved them to read whenever I needed a dose of inspiration. And now, she knew of me. The column was titled "Celebrating the Bold Thinking of ASHA Imaginologists." My eyes were glued to the screen:

> I cannot think of a better use for my last column as ASHA's president than to highlight the stories of some of our colleagues who have dared to push their personal boundaries beyond the safe and familiar. While each is unique, each began with the seed of an idea, that fertilized with equal measures of passion, vision, and hard work, eventually blossomed into reality. I hope the four stories I have chosen to share here will delight, inspire, and challenge you to imagine more and imagine better.[42]

In the next paragraph, entitled "A hunger for words," Dr. Robertson wrote about the work I had done with Stella. I truly could not believe it. I had completely forgotten that a few months ago I submitted a blurb about Stella's communication journey to ASHA for a chance to be featured on their social media accounts. I never dreamed the president would see it and end up writing about it.

This news could not have come at a more perfect time. I was excited for my phone interview with *People* the next day, but ever since I created my website, I was a little nervous about what ASHA would think. I was proud of my career and my profession. All I wanted was to represent the field of speech-language pathology as best as I could. Now, seeing such enthusiastic support from the president of ASHA, I was fueled to keep going and to keep sharing. It felt like a permission slip to continue my pursuit.

November 4, 2019, started out like any other Monday. I listened to an inspiring podcast while I drank a cup of coffee, had three speech therapy sessions, and stopped home for lunch. But, instead of eating lunch and taking Stella for a quick walk, I had a phone interview with Hilary from *People* magazine. Unlike my interview with the NIU alumni magazine writer, I had no idea what the questions would be.

"Are you sure you're ready for this?" she asked. "Millions of people will see it. And I just want to warn you, there are some really mean people who will find something negative to say about even the most positive stories."

I took a deep breath. "I'm ready," I said. "This is too important not to share, regardless of what anybody else thinks."

Hilary told me she didn't know when the article would run. "Sometimes it's a couple of days, and sometimes it's a couple of weeks. I'll send you the link when it's up though."

Forty-five minutes later, I was off the phone and on the road again to my next client. I received a last-minute text from the parent of my 3:00 session saying they had to cancel. So this was my last session of the day.

I could not stop running everything I said back through my head. Except I could not even remember what I said. It was such a blur. I hope I represented what I was doing well enough. I hope I explained the concept in a way that everyone would understand. Thankfully, my job is not one where I can keep ruminating on something in my head while continuing to work. Toddlers require full attention and truly keep you in the moment. Their stream of energy, curiosity, and wonder about everything around them, and desire to play is magnetic. There was nothing else for me to do now except to play, laugh, have fun, and, my favorite thing of all, teach more words.

At the end of the session we sang the cleanup song and I pulled my phone out of my bag to write my session note. My screen was filled with notifications from Gmail and Instagram, and a text from Hilary. *She probably has a follow-up question*, I thought.

But I unlocked my phone to see the words It's up! with a link to an article headlined "Dog Learning to Talk By Using a Custom Soundboard." I quickly scrolled down to see pictures of Stella with her AAC device, and the two of us together. I had not mentally prepared at all for this. I had no idea it would be posted so quickly. I was so focused on getting through the interview that I did not even think about what might happen when it was online. My heart pounded. I so badly wanted to read it. But I was sitting on the floor of a client's living room, needing to write my session note so his mom could sign it.

As soon as I shut the front door behind me, I clicked on the article link again. I sat in my driver's seat not believing that there were pictures of Stella and me in a *People* article. Immediately, I texted it to Jake, my family, and friends. I opened Instagram to see that I already had a couple thousand new followers in the hour since it was posted. Jake picked up champagne on his way home. We toasted to a growing community of fans and my work officially being "out there."

Before we poured a single glass of champagne, my in-box was rapidly filling up with questions, pictures of dogs from passionate new followers, usage permissions for my videos from random news outlets, and all kinds of media requests. I had no idea that other news outlets would want to write about this story once they saw it in *People*, and I had no idea this would all happen so quickly.

Inside Edition wanted to come film a segment at our apartment. They said they needed an answer right then because they move on to new stories pretty quickly. I didn't want to miss an opportunity, so I said yes. I refreshed my in-box to see that *Popular Mechanics* asked to interview me the next day, and CNN wanted to conduct a Skype interview the day after that. I couldn't even drink a glass of champagne. I had to figure out what to do. I still had a job. I had a full week of kids to see. Was I supposed to give these people permission to use my videos? I had absolutely no idea. How many interviews should I do? Would these requests be the only ones I would ever be asked to do, or would they still be interested in the future? Nobody I knew had been through anything like this.

For the rest of the night, Jake and I grew dizzy as we watched my number of followers keep jumping, new spin-off articles being posted rapidly, and the story bouncing through Twitter and Reddit.

You're trending on Apple News! my sister texted our family group chat. We all took guesses on how many followers I would reach by the end of the night, and my dad kept sending screenshots of heartwarming comments on my Instagram videos. I figured this was going to be the peak, that in the next day or two it would all slow back down again, and these would be a fun couple of nights to remember. When we went to bed, I was at about ten thousand followers. I never imagined I would go from six hundred to ten thousand in one single day.

The next morning, I woke up to thirty thousand followers. People were now commenting in different languages. A text from my cousin Rachel read: I had the news on this morning and all of a sudden they were talking about you!! She attached a video clip of *CBS This Morning*. Gayle King was talking about Stella and me. A picture of Stella's face took up the entire backdrop behind the anchors, and one of my Instagram videos of Stella talking played. Random acquaintances from middle school through college were coming out of the woodwork, messaging me screenshots of people they knew posting articles about me. This was completely wild. "I can't believe I have a session with a two-year-old in an hour," I said. "How in the world am I supposed to focus?"

In between therapy sessions, I pulled into a gas station parking lot and had a phone interview with *Popular Mechanics*. Then, I scheduled *NBC Nightly News*'s visit to our apartment.

Parents of the children I worked with were seeing articles about me. When I showed up at the front door of a client's house ready for therapy, the mom opened the door. "There's the face I've been seeing all over the news," she said. "Someone showed me a Spanish news clip about a talking dog, and I screamed when I saw it was you!"

I often pictured how my work with Stella would become known to the public. I thought I'd gain a slow but steady following. I thought it would catch the attention of someone in the dog world and that person would pass it around to other people in the dog field. That's what had started to happen with the speech therapy community. I pictured myself teaching several different dogs before it really gained the attention of the public. But I truly believed this would not stay hidden forever. Stella's progress was all too incredible. I did not know exactly what would happen, or when, but I felt somewhere deep in my core that this would all come out in some way, shape, or form.

All the threads that came together at the right time in the right way astounded me. Sending my website to my former professors led to the NIU Alumni Association reaching out for an interview. The article in the university magazine attracted a lot of people in the area, especially speech-language pathologists. One speech therapist who saw that article shared one of my videos of Stella on Facebook. Her cousin, a writer for *People*, saw the video and reached out to me. Every single small step I took in moving my project forward was necessary for it to be shared with the world.

Thursday morning an *Inside Edition* camera crew from L.A. parked on the street outside of our apartment. Two men in their late thirties stepped out of the van. Stella immediately ran up to the interviewer with her typical body wiggle and smile. He stood near the curb with his arms crossed, looking down to her. "I didn't think she would be energetic," he said.

"Oh, yeah. She's only one . . . she's just excited," I said. "She loves people." He patted her head once, then did not interact with her again.

Jake and I exchanged glances that said, *I sure hope this was a good idea*.

I was incredibly nervous. I had no idea how Stella would react to cameras or if she would talk with complete strangers and her space overtaken by video equipment. I tried keeping Stella entertained and comfortable while they set up their equipment in our living room. The cameraman squatted down to Stella's level while carrying a three-foot-long, bulky camera on his shoulder. He trailed behind her, following her every move. Stella looked back at the camera and trotted over to her board.

"All done," she said. Stella ran into the bedroom and hid behind the bed.

"Let's do your interview first," he said. "We'll give her some time to get used to us."

Over the two hours that they stayed at our apartment, Stella said several different words. She said when she wanted to go to the park, play with her ball, she was mad that we could not go outside again, and said "bye" when they started packing up. I was so relieved that Stella adjusted well and that they could capture footage of her talking. I could not wait to see what they would use when they ran the segment.

Two hours later, I signed onto Skype for an interview with *CNN International*. I had no idea who was interviewing me, or what they looked like. I had to stare at a black screen while the two anchors talked between themselves and asked me questions. It was strange being unable to see what was happening, and knowing that they, along with all the viewers, could see me.

"I don't think I can ever watch that," I said to Jake. "I don't even want to know what I looked like or said."

When the *NBC* crew arrived from New York and Los Angeles that Friday, I felt a little less nervous now that I knew Stella would probably react well. I already had two interviews under my belt and knew what types of questions to expect. The *CNN* segment had aired and made its way around. They put together an excellent compilation of videos of Stella talking, the interview, and pictures. It turned out much better than I expected it would.

Instead of only two people, the *NBC* crew consisted of five. They brought in more equipment than I thought could fit in our entire apartment. We rearranged some of our living room furniture to make it all fit. Lighting umbrellas surrounded Stella's board; cameras were set up at every possible angle. We could all barely squeeze into our small space. Once they finished preparing the setup, everyone sat in a row on our chaise lounger, dining stools, and living room chairs staring at Stella. Stella looked at everyone and walked over to her board.

"Hi," she said. She wagged her tail and walked over to greet them again. Stella must have been so confused. Suddenly there were so many new people coming over and putting all kinds of strange objects around her board.

Inside Edition had asked me several questions about language, AAC, and Stella's progress. I was disappointed when they aired the segment, though. They did not include any of my responses. The quotes they chose from me were, "Yeah, it's crazy" and me answering the question about what her least favorite words were. It was worse with *NBC.* They spent over five hours at our apartment. I took the day off from working with my kids. We even all went to the dog beach together. After all that, they did not end up airing the segment. It was due to the news cycle. I had no idea it was not a guarantee. Still, the potential of the breakthroughs I had made with Stella was clearer than ever. The writer from *People* left a voice mail saying that article was one of the most well-received stories they had ever published. Thousands of kind people emailed me asking to learn more, or to share stories about what their dogs understood.

Two weeks after the *People* article came out, I had over five hundred thousand followers on Instagram. My posts were flooded with comments and questions. My parents put together a little booklet of some touching comments for me to look at when I needed to remember the bigger picture of what I was doing. It was such an outpouring of support, excitement, enthusiasm, and joy. I felt like the world was ready for my work. It was time for me to figure out how to share even more of it, and how we reached this point.

Even though it was thrilling to see the great interest, I quickly came to see that it would not matter to me what anyone else thought about what I was doing. Stella's accomplishments were all that mattered to me.

This quote from Deepak Chopra struck me during this time of extremely heightened visibility: "Happiness for a reason is just another form of misery because the reason can be taken away from us at any time."

I did not want my feelings toward my work to be contingent on anyone else's perceptions of it. I did not want to start the habit of saying, "There are so many positive comments on this video, I'm so happy," then seeing a negative comment and being upset by it. I wanted to feel good about the work, and free myself from the approval or disapproval of an internet stranger. I started the habit of checking in with myself to ask myself if I was happy with what I was doing and with Stella's progress. Or I would ask myself, *If I didn't have a bunch of people watching me, what would I do?* Then I would proceed to do that.

The media requests and emails still poured in. Somehow, Jake and I were both receiving phone calls from international radio stations asking to speak with me. A French station repeatedly called and texted Jake. Radio stations called my parents' home in Illinois. I still have no idea how anyone found our numbers. A Canadian news station called the speech therapy company I worked for, looking to talk to me. Local news anchors sent me friend requests to try making contact. It was all complete chaos.

At one point, I turned off the lights in my bedroom and pulled the sheets over my head. Jake rubbed my back. "It feels like hundreds of thousands of people are tugging on my shirt all at once, asking me for things," I said. I still had kids to see and a job to do. I desperately wanted to teach people, but I could not explain speech therapy and a year and a half full of observations in Instagram comments or in thousands of email replies. If I tried responding to everything, I would not have time to do anything else in the day. I needed to figure out what I wanted. "How am I supposed to manage all this?" I asked. If I had had any idea this was going to be so big, I would have tried to prepare myself first.

I soon realized that the depth of this story could not be reflected in the quick news cycles. I would not be able to communicate my patterns of thinking as a speech therapist and the past year and a half of progress and observations in minute-long video clips. It was not even up to me to choose which of my lines were used or not. I wanted to be able to tell our story and help people all over the world see dog communication and potential through the lens of a speech-language pathologist. And I did not want to leave it up to anyone else to share my work, this exciting new idea, for the first time.

I had offers to do absolutely everything I could think of: speaking engagements, TV show ideas, interviews, documentaries, product creations, course creations, and one subject line of an email that caught my attention the most: "Write a book?"

CHAPTER 18

This Is the Beginning

After the media craziness died down, I filled up the rest of Stella's board and continued teaching her new words. With the three new additions, *blanket*, *like*, *where*, we reached thirty-two buttons. The first time Stella used "like" was when we found one of her favorite balls that had been lost under the couch for months. When we rolled it out, Stella picked up her ball, said "ball like," and continued playing with it for hours.

I added "where" to see if Stella would be able to ask us questions. "Where" is one of the first question words that toddlers start using. "Where Mom?" "Where Dad?" Where ball?" are all common phrases for toddlers. I modeled "where" every time I asked Stella where she wanted to go, or where one of her toys was. After only a few days of modeling, when Jake and I would put our shoes on, she asked "where?" When we would return from the store, she sniffed the bags and asked "where?" When I was talking to Jake on speakerphone, Stella asked "where?" and looked out the window. When her toys or blanket were out of sight, she would ask, "blanket where?" or "where toy?" When I came home from work one day, I asked, "Want to go for a walk?"

Stella paused for a few moments. "Where?" she asked. She looked back up to me.

"Walk on beach," I said.

Stella wagged her tail and said, "Walk outside."

She cared about where we were going and what we were doing. I wondered how long she had been wanting to ask us "where," after we had asked her "Where do you want to go?" for months. What else was Stella wondering? What else was she waiting to be able to say? I guess I would need more space and more buttons to find out.

So what happened when I used speech therapy techniques to teach words to my puppy? When Stella was eight weeks old, I introduced just one button that said "outside." My initial plans were to give Stella a way to say a few different words so we would know what she wanted. I recognized all the other ways she was communicating with gestures and vocalizations, how she was understanding words, then continued to add more and more vocabulary as we progressed together. My vision became clearer with each step forward I took. As Martin Luther King Jr. said, "Faith is taking the first step even when you don't see the whole staircase."

Now, Stella uses nouns, verbs, names, adjectives, and question words to tell me what she wants to do, where she wants to go, when she is thinking about me, what we are doing, what she likes, when she is mad, when she is happy, when she needs alone time, when we are being good, to answer questions, to ask questions, to participate in short conversations, and to make her own unique phrases every single day. My process of teaching Stella language certainly was not perfect. It took several months before I even had the idea to put all her words together on one cohesive board for her. I hope dogs of the future will continue reaching new milestones that even Stella has not achieved yet. This is just the start. It is the door opening to show what is possible, and all that we as a society have to explore and learn.

The question I am asked the most is "How did you come up with this idea?" The easiest explanation is that I worked with lots of children who use communication devices to talk. I recognized prelinguistic skills in Stella's communication that toddlers demonstrate before they start saying words. So I wondered if Stella could say words if she had a communication device.

But the real reason is so much deeper than my experience as a speech-language pathologist. I had no preconceived notions of what could or could not be achieved. I had never spent time reading about why dogs cannot talk, or why animals will never use human words. I was completely dedicated to my field of speech therapy and saw a fun opportunity to try it out in a new way. I saw all the reasons why it could be possible. I looked for the possibilities rather than the potential problems. I put faith into my own professional experiences rather than looking at others' expertise as the gold standard. I put weight into my own ideas instead of what the world thought was possible. I did not have any limiting beliefs stopping me from discovering more.

What would have happened if Thomas Edison gave up on his experiments because someone else did not share his vision?

What would have happened if the Wright Brothers allowed other people's ideas of what was possible to stop them from trying something new? All great ideas and discoveries come not from focusing on "what is," but from imagining what "could be." After all, as Albert Einstein said, "Logic will take you from A to B. Imagination will take you everywhere." I choose to live in a world where visions of what "could be" are celebrated and prioritized.

Language is often viewed as the last barrier we have from the human to the animal world. What happens when that barrier dissolves? We realize that we are all connected in more ways than we can even begin to comprehend. We are not as separate from animals as many think we are. We all think, we all feel, we all have opinions, we all communicate, and we all want to connect.

I am a firm believer that our beliefs about the world shape our own individual experiences. I challenge you to think in terms of possibilities instead of problems. Ask yourself, *What if it does work? What if the results are even better than I thought?* I challenge you to look at the world through the lens of untapped potential everywhere you turn. I challenge you to follow your curiosities, no matter how wild or unlikely they may seem at first. I challenge you to let go of all the excuses you can think of for why something couldn't work and hold on to the reasons why it could. At the start of any pursuit, you have no idea what it could become.

Starting from the first days I spent with little Stella, I watched her progress through stages of language development in ways extremely similar to human children. When I gave Stella the opportunity to use words, she continually pushed the envelope and acquired new skills beyond my expectations. She always found ways to use words in unique situations, make longer phrases, and communicate about new experiences. On average, Stella said between twenty-five and forty different utterances each day. Sometimes, she even said between fifty and sixty different words or phrases on a given day. She had so much to say and so badly wanted to communicate with us about the world we were all living in together.

I thought bringing home a puppy would change my life in all the predictable ways. I expected to play with Stella, and to cuddle with her. I expected to go for more walks, to watch her grow, to care for her, and to learn about her personality. I expected to have a companion who would stay by Jake's and my side as we began our life together.

But Stella gave me so much more than this. She completely changed my life. I can never go back to the way I looked at the world before Stella learned to talk. Stella opened my eyes to the glaring similarities between dog and human communication skills. She made me spend my spare time contemplating humans' relationships with the animal world. She made me never question a single child's communication potential ever again. She made me work my hardest to get AAC devices for children. She taught me to make the most out of every window of opportunity. She taught me how powerful believing in potential can be. She taught me that everyone deserves a chance to use their voice and that everyone has something important to say. She made me realize how desperately I wanted to open the world's eyes to the power of communication. She introduced me to my life's purpose.

Stella's and my journey represents the beginning of a field bursting at the seams with possibility. We have so much to experiment with, so many questions to ask, and so many ideas to test out. What other words can dogs learn? What is the range of normal for dogs' language skills? Is there a critical period for language learning? How long would it take older dogs to learn to use words in comparison to puppies? Do dog breeds differ in language capabilities? What are dogs' collective syntax patterns? How do device setups impact language use? Will service dogs be able to help their owners more if they can communicate with words? How can we adapt devices for dogs with special needs? How does having access to words change a dog's awareness and perception of their environment? Will we stop assuming that we know everything our pets are thinking? Will we stop talking for our animals and instead let them speak up for themselves?

Will this open new doors for interspecies communication research? What animals could be next? How will our new communication with dogs change how we treat them? How will this change our pets' lives? How will this change ours?

It is time for us to start searching for the answers to these questions and many more. It is time to unite the human and animal worlds through language. It is time for a communication revolution. And I cannot wait to see what we all discover together.

APPENDIX A

Helping Your Dog Learn to Talk

Talk to your dog!

All dogs, regardless of their age, are bursting with communication potential. Narrate your dog's actions and activities in short, simple phrases. If you are not sure what to say, ask yourself, *What is happening right now?* Examples of narration include "Stella eat," "Play toy," "Stella Christina walk," "Water!"

Observe your dog's nonverbal and verbal communication

How does your dog currently communicate? Does she paw at objects? Does she whine when trying to grab your attention? Does her tail wag when she is happy? Does she bark at you when she needs something? Notice all the ways your dog communicates and start responding. We want to respond to all communication, not only to words. The more we acknowledge all forms of communication, the more likely your dog will be to communicate with words.

Try modeling words when your dog is communicating verbally or nonverbally. When you see your dog pawing at the door, say "outside." If your dog whines near her leash, say "walk." If your dog looks at a toy that is stuck under the couch, say "help." Pairing words with gestures or vocalizations that your dog already uses helps her learn the meanings of the words.

Pay attention to words your dog is understanding

Which words cause your dog to become excited, turn her head back and forth, or run in from the other room? Are there any words you have to spell out, so your dog does not hear what you are planning? Keep track of these words. Dogs will likely learn to say words that they already know the fastest.

Select words to teach

Ask yourself the following questions: Which words is my dog already understanding? What is my dog already communicating to me with gestures and vocalizations? Which words are frequently occurring? Which words would help her communicate about a variety of experiences? Which words would be most beneficial for all of us if she could say them? What does my dog love to do? Use the answers to these questions to guide your decisions. Start with words that your dog will want to say and will have plenty of opportunities to say.

Set your dog's device up in a convenient location that is easy to access for you and your dog

Find a space in your home where your dog naturally spends time already. Try picking a spot that has room to support potential growth and that will stay clear of clutter covering or hiding the buttons.

Keep talking to your dog, but push the corresponding buttons when you talk

Keep narrating your dog's actions but use your dog's buttons to talk as well! This is called *aided language input*. The more we use communication devices as we are talking, the more AAC users learn how to use their device to say words. Before you take your dog outside, say "outside" verbally and with her button. For even more learning, set a goal for yourself to repeat a single word between five and ten times before moving on. Refer to Chapter 5 for an example of how to incorporate this into your routines.

Pay attention for signs that your dog is observing your modeling

It might take time for your dog to notice the buttons or device, or to really pay attention to you modeling words. Even if it does not look like your dog is watching, keep on modeling. She is still hearing the words you are saying. Look for signs that your dog might be becoming aware of the device. Does your dog look at your foot, or up to you? Does your dog glance down at the button while walking past? These are all signs that your dog is headed in the direction of exploring the words.

When your dog is becoming more aware of your modeling, create communication opportunities within your routines

When you see your dog noticing your modeling or noticing the buttons, turn your routine interactions into language-facilitating opportunities. The greatest cue we can provide is a long, silent pause to give the AAC user a chance to process what is happening and try exploring her words. When you see your dog communicate through a gesture or vocalization, stay quiet for at least ten to fifteen seconds. At the end of fifteen seconds, if your dog looks like she might be walking toward her buttons or is looking at them, continue staying quiet. If you have not seen an indication that she might try saying a word, add the next level of naturalistic cueing.

Add naturalistic cues if necessary

After you have already tried staying silent, you can try standing next to your dog's buttons or pointing at them. Dogs respond to human gestural social cues as early as six weeks old! If pointing or standing near the buttons does not work, you can try tapping it with your finger or foot, not pushing it to activate it. You can also try adding an open-ended verbal prompt such as, "What do you want?" Avoid telling your dog exactly what to say or forcing your dog to push a button with their paw. This can cause prompt dependency and keep your dog from learning how to use words independently.

If none of these prompts encourage your dog to try saying a word, simply model it again verbally and with the button, and carry on with the appropriate action.

Continue responding to all forms of communication

Do not withhold food, water, trips outside, playtime, or anything from your dog until she says a word. Respond to all forms of communication, and allow for a minute or two of opportunity for your dog to try using a word.

Provide positive praise when your dog tries to use their device

If your dog approaches the device, sniffs it, paws at it, licks it, stares at it, be excited! Show your dog that you are proud that she is exploring and testing out the new object in her space. If your dog successfully pushes a button, respond to it! Even if you think it was accidental or she was just pushing it to explore, respond! Your dog will learn the meaning of the words by observing your reactions whenever she uses it.

Respond to words

Especially while your dog is in the beginning stages of learning, respond to your dog's words as frequently and as best you can. Again, even if you do not yet know if they are intentional words, keep responding as if they are. Providing the appropriate response will help your dog learn how to use each word intentionally. If your dog says a word, but you think she really meant to say something else, still respond to the word your dog said. This provides a great opportunity for your dog to distinguish between words. After you respond to the original word, you can model the word you thought your dog was wanting to say.

Keep modeling

When your dog starts using words, do not stop modeling! The more you use your dog's device, the more she will continue to learn and use it on her own. Try to model words in a variety of contexts. This will help your dog generalize the meaning of words to multiple situations and show her the many different ways she can use it.

Keep adding vocabulary that will allow your dog to talk about a variety of experiences

When in doubt, try adding more words! You might be surprised by how much your dog learns to say or what she prefers to talk about. Think again about the words you use often when talking to your dog, her daily activities, and different functions of communication. Does your dog have words that she could use for communication functions other than requesting? Does she have a way to tell you "no" or "all done"? Can she call you over, or let you know when she is happy or mad?

Model two-word phrases when your dog starts using single words

As your dog reaches new milestones, keep modeling the next level up. Model short phrases such as "play outside," "walk outside," "come play." If you are finding that the vocabulary you have makes it challenging to combine words, try adding more verbs instead of nouns. If it is difficult for you to combine words with your dog's vocabulary, it will be even more challenging for your dog.

Have fun!

Teaching your dog words, learning more about her thoughts and personality, and connecting on a deeper level is incredibly rewarding. Enjoy the process and be proud of any progress that you and your dog make together.

Troubleshooting

What to do if . . .

You are wondering how to teach a specific word

Think about your own communication. In which contexts do *you* naturally use this word? These are the situations in which you should model it. For example, "How do I use the word *want*?" I use *want* when I am requesting an object or activity.

So every time I see Stella requesting an object or activity, I will model *want* to go along with her words or gestures.

Your dog says a word that you were not expecting to hear or that seems "random"

Respond to it! Your dog may be exploring new vocabulary. The only way she will learn the meaning of each word is by seeing your response to it. If after you respond, your dog seems to not have received what she expected, model the word you think she meant to say, then follow through with that natural response.

Your dog is scared of the buttons or device

Don't force your dog to try using the buttons or be near them. Keep the buttons out for a few days before you try using them again. Spend time sitting or standing by the buttons so your dog sees that they are not harmful. If possible, call your dog over to pet her while near the buttons. When your dog is calm, try modeling a word again and providing the appropriate response. The calmer you are, the calmer your dog will feel.

Your dog only says words on your command, not spontaneously

Stop using the verbal prompts you are giving immediately. Your dog has learned the pattern of waiting for you to give a cue, then saying a word rather than using the buttons on her own. Instead of asking "Do you want to go outside?" or saying, "Tell me outside," simply say "outside" verbally and with the button before taking your dog out. When you notice your dog gesturing or vocalizing to go outside, provide subtle, naturalistic cues. Start with silently standing near the button or looking at it. Increase to pointing at it or tapping it if necessary.

Remember to wait at least ten to fifteen seconds before adding a cue. Sometimes a long pause is all that is needed.

Your dog continuously requests the same object or activity

If your dog has a solid understanding of the word for the object or activity, incorporate an "all done," "finished," "stop," or "later" button. You can respond by saying "all done walk" or "walk later" both verbally and with the buttons. Provide a suggestion for what your dog could do now instead: "all done walk, play now!" then start playing. Or "all done walk, bed now" and pat your dog's bed.

Also, consider the amount of vocabulary you have available for your dog. If your dog only has a few words, think about if your dog might be using one word to communicate several different wants or needs. Providing access to more words helps your dog differentiate their different meanings.

Your dog is not using the buttons after you have been modeling them for a few weeks

Look for subtle changes in your dog's progress. Is your dog stopping to look at the buttons? Is she walking slower past them? Is she watching you model? Is she standing by the buttons? All these are steps in the right direction. Strive for these stages before striving to hear your dog's first words. Progress can take time, and steps forward can be subtle.

You can also try adding and modeling new words. Your dog might be more excited to say a different word than you thought she would be!

You need to switch your dog's setup

Understand that changing devices will take time for your dog to relearn the locations of her words. If possible, keep the old device out and available for at least the first couple of days, and model words on the new device. This can be less traumatic than completely removing a familiar setup. Help your dog through this transition by providing high-frequency models on the new device, being aware of all your dog's forms of communication, and providing lots of positive praise and encouragement when your dog explores the new device.

If You're Not Seeing the Progress You Expected, Ask Yourself:

Are your dog's basic needs being met?

Is she feeling safe and secure in her environment? Is she well rested? Is she getting enough playtime? How your dog is feeling impacts her ability to learn. If she is stressed, tired, sick, scared, or overwhelmed, it will be more difficult for her to learn new information.

Is the environment chaotic when you're modeling?

It might be difficult for your dog to concentrate if there are a lot of other noises or activities happening. Try modeling words and working with your dog when the environment is calmer.

Is everyone in the household modeling?

The more people who use your dog's communication device to talk, the better your dog will learn. Consistency is important!

Are you providing a long pause before you give another cue?

Before jumping in to use one of your dog's buttons, wait at least ten to fifteen seconds. AAC users and emerging communicators benefit from time to think and process what is happening.

Are the words motivating for your dog?

Make sure the first words you select are words that you think your dog would want to say! Think about what your dog already communicates via gestures and vocalizations, and which words she is excited to hear you say.

Do you have too few words available?

Having more vocabulary available to model and incorporate into your routines can help your dog learn faster. Greater exposure leads to greater learning.

Are you directing your dog to talk on command?

Telling your dog what to say and when can hinder their ability to use words independently. They learn to say what you tell them instead of what they want to say. The ultimate goal is spontaneous communication with words, not talking on command. Instead of giving your dog a command to say a specific word, model each word in the appropriate contexts.

Is your device in an easy-to-access location for your dog?

Make sure to keep your dog's buttons or device in a room where she naturally spends a lot of time.

Are you keeping your buttons in the same location?

Moving word locations around can be confusing for your dog. This would be the equivalent of switching the location of the keys on the keyboard every time you typed. Keeping the words in the same spots will lead to the fastest learning.

APPENDIX B

Resources and Recommended Reading

To learn more about augmentative and alternative communication, language development, or speech therapy, visit:

Hunger for Words: www.hungerforwords.com

Hunger for Words was created by Christina Hunger to share her communication journey with Stella, educate readers about speech therapy and AAC, and inspire others to teach their pets to talk. You can find videos of Stella talking, answers to frequently asked questions, and resources to help you on your journey with your dog.

The American Speech and Hearing Association (ASHA): www.asha.org

The American Speech-Language-Hearing Association (ASHA) is the national professional, scientific, and credentialing association for audiologists and speech-language pathologists. ASHA provides information about the speech therapy scope of practice, research materials, and evidence-based interventions.

AssistiveWare: www.assistiveware.com

AssistiveWare is an assistive technology company whose website includes a multitude of easy-to-read articles about important AAC topics. Their "Learn AAC" section is an excellent place to start if you are looking to read about presuming competence, modeling, different functions of communication, and how to be a good communication partner.

Language Acquisition through Motor Planning (LAMP): www.aacandautism.com/lamp

LAMP is a therapy approach developed by a speech-language pathologist and an occupational therapist. Based on the principles of motor learning and natural language development, the LAMP approach is highly effective in teaching words and improving functional communication. The LAMP website shares information about the power of learning words through motor planning and important AAC device features.

PrAACtical AAC: www.praacticalaac.org

PrAACtical AAC is a blog created by AAC researcher and professor Carole Zangari. PrAACtical AAC has a wealth of therapy tips and strategies for helping AAC users become effective communicators.

AAC Language Lab: www.aaclanguagelab.com/language-stages

Created by one of the leading AAC device manufacturers, PRC, the AAC Language Lab provides support materials for speech-language pathologists, teachers, and parents of AAC users. Their "Language Stages" gives excellent descriptions of each phase of language development, and objectives to target during each one.

The Hanen Centre: www.hanen.org

The Hanen Centre, a nonprofit founded by a speech-language pathologist, trains parents on how to help their young children develop language and communication skills. They offer courses, guidebooks, workshops, and several free articles about effective language facilitation strategies.

Recommended Reading

Thirty Million Words: Building a Child's Brain by Dana L. Suskind

Chasing Doctor Dolittle: Learning the Language of Animals by Con Slobodchikoff

Chaser: Unlocking the Genius of the Dog Who Knows a Thousand Words by John Pilley and Hilary Hinzmann

The Inner Life of Animals by Peter Wohllben

The Education of Koko by Francine Patterson

The Genius of Dogs by Brian Hare and Vanessa Woods

The Soul of an Octopus: A Surprising Exploration into the Wonder of Consciousness by Sy Montgomery

Acknowledgments

To my parents, Laura and Brian Hunger, thank you for raising me to be the person that I am today. You taught me from such a young age that my voice and opinions matter, and to speak up for what I believe in. Thank you for encouraging me to think differently from the crowd and for your constant love and support.

To my big sisters, Sarah Hunger and Kate Elliott, thank you for being such incredible role models for me while growing up. You both trailblazed your own paths and inspired me to create my own as well. You have no idea how much I wanted to grow up to be like you two.

To my great friend, Grace Stevens, I truly believe that our meeting and working together was written in the stars. Thank you for always giving such sound advice about both speech therapy and life, brainstorming ideas about Stella's AAC with me, reading drafts of my book, and being a fantastic friend.

To my wonderful friend Sarah Reece, thank you for coming up with the name Hunger for Words and for all of our AAC-inspired discussions. You are quite possibly the only person I know who would remind me to bring my iPad whenever we hung out so we could talk to each other using AAC. All of your students are so lucky to have you as their speech therapist.

To the entire Northern Illinois University Speech-Language Pathology department, thank you for equipping me with an excellent education and for introducing me to the fascinating world of augmentative communication. And thank you to Michelle O'Loughlin, my first clinical supervisor, mentor, and now great friend. You taught me the importance of listening to myself and thinking outside of the box.

To my literary agents, Christopher Hermelin and Ryan Fischer-Harbage, I am so grateful for all of your ideas, encouragement, and enthusiasm throughout my entire book-writing process. Christopher, you have been such a wonderful literary agent, sounding board, and friend to me. Thank you for living in the land of possibilities with me and for helping me share my voice with the world. Ryan, thank you so much for your devotion to my book. No matter what was going on, I could always count on you to give such excellent advice. Thank you both for introducing me to this next phase of my life.

To my editor, Mauro DiPreta, thank you for believing in my potential to write this book. I have learned so much from working with you and am incredibly grateful for all of your comments, suggestions, and editorial visions. You helped me turn my ideas for *How Stella Learned to Talk* into a wonderful reality.

To the entire William Morrow team, thank you so much for all of your hard work in bringing my book to life and sharing it with the world. Tavia, Jamie, Vedika, and Kelly, it has been so fun getting to know you and working with you all. I could not have asked for a better group of people to guide me through this process and work on this book.

To my Hunger for Words community and to all of my readers, thank you so much for being a part of this journey. I can't believe how far this movement is spreading, and that is all thanks to you. I am so unbelievably proud seeing how many of you are using AAC with your own dogs (and other pets!). You are all an important part of this new era of interspecies communication, and are showing the world that everyone has a voice.

To my husband, Jake, to whom this book is dedicated, words cannot begin to describe how thankful I am for your unconditional love and support. You have believed in me every single day through every single pursuit, and it has truly meant the world to me. Thank you for teaching Stella with me, reading each draft of my book multiple times, and for being the biggest supporter of all my dreams. I can't wait to see what we do together next.

And finally, thank you to Stella. I knew from the moment we met that we were in for a special journey together, but this is wilder than I could have ever imagined. Thank you, sweet girl, for your kisses, body wiggles, smiles, and love. You are my inspiration. I love you so much.

Notes

1. "Catahoula Leopard Dog," DogTime, accessed September 9, 2020. https://dogtime.com/dog-breeds/catahoula-leopard-dog.
2. "Australian Cattle Dog Breed Information, Pictures, Characteristics & Facts," DogTime, accessed September 9, 2020. https://dogtime.com/dog-breeds/australian-cattle-dog.
3. Jana M. Iverson and Susan Goldin-Meadow, "Gesture Paves the Way for Language Development," *Psychological Science* 16, no. 5 (May 1, 2005): 367–71. https://doi.org/10.1111/j.0956-7976.2005.01542.x.
4. Louis Michael Rossetti, *The Rossetti Infant Toddler Language Scale: A Measure of Communication and Interaction* (Austin, TX: PRO-ED, Inc., 2005).
5. Rossetti, *The Rossetti Infant Toddler Language Scale.*
6. Janet R. Lanza and Lynn K. Flahive, *Guide to Communication Milestones* (East Moline, IL: LinguiSystems, 2008).
7. Gregg Vanderheiden, "A Journey Through Early Augmentative Communication and Computer Access," *Journal of Rehabilitation Research and Development* 39, no. 6 (2002): 39–53.
8. John W. Pilley and Alliston K. Reid, "Border Collie Comprehends Object Names as Verbal Referents," *Behavioural Processes* 86, no. 2 (2011): 184–95. https://doi.org/10.1016/j.beproc.2010.11.007.

9. John W. Pilley, "Border Collie Comprehends Sentences Containing a Prepositional Object, Verb, and Direct Object," *Learning and Motivation* 44, no. 4 (November 1, 2013): 229–40. https://doi.org/10.1016/j.lmot.2013.02.003.
10. Attila Andics, Anna Gábor, Márta Gácsi, Tamás Faragó, Dora Szabo, and Adam Miklosi, "Neural Mechanisms for Lexical Processing in Dogs," *Science* 353, no. 6303 (September 2, 2016): 1030–32. https://doi.org/10.1126/science.aan3276.
11. Nell Greenfieldboyce, "Their Masters' Voices: Dogs Understand Tone and Meaning of Words," NPR, August 30, 2016, www.npr.org/sections/health-shots/2016/08/30/491935800/their-masters-voices-dogs-understand-tone-and-meaning-of-words.
12. Tara O'Neill, Janice Light, and Lauramarie Pope, "Effects of Interventions That Include Aided Augmentative and Alternative Communication Input on the Communication of Individuals with Complex Communication Needs: A Meta-Analysis," *Journal of Speech, Language, and Hearing Research* 61, no. 7 (July 13, 2018): 1743–65. https://doi.org/10.1044/2018_jslhr-l-17-0132.
13. Luigi Girolametto, Patsy Steig Pearce, and Elaine Weitzman, "Interactive Focused Stimulation for Toddlers with Expressive Vocabulary Delays," *Journal of Speech and Hearing Research* 39, no. 6 (December 1, 1996): 1274–83. https://doi.org/10.1044/jshr.3906.1274.
14. Shakila Dada and Erna Alant, "The Effect of Aided Language Stimulation on Vocabulary Acquisition in Children with Little or No Functional Speech," *American Journal of Speech-Language Pathology* 18, no. 1 (February 1, 2009): 50–64. https://doi.org/10.1044/1058-0360(2008/07-0018).
15. Napolean Hill, *Think and Grow Rich* (New York: Jeremy P. Tarcher/Penguin, 2005).
16. Hill, *Think and Grow Rich.*
17. Erinn H. Finke, Jennifer M. Davis, Morgan Benedict, Lauren Goga, Jennifer Kelly, Lauren Palumbo, Tanika Peart, and Samantha Waters, "Effects of a Least-to-Most Prompting Procedure on Multisymbol Message Production in Children with Autism Spectrum Disorder Who Use Augmentative and Alternative Communication," *American Journal of Speech-Language Pathology* 26, no. 1 (2017): 81–98. https://doi.org/10.1044/2016_ajslp-14-0187.
18. Hilary Mathis, "The Effect of Pause Time Upon the Communicative Interactions of Young People Who Use Augmentative and Al-

ternative Communication," *International Journal of Speech-Language Pathology* 13, no. 5 (2011): 411–21. https://doi.org/10.3109/17549507.2011.524709.

19. "Dogs' Intelligence on Par with Two-Year-Old Human, Canine Researcher Says," American Psychological Association, August 10, 2009. https://www.apa.org/news/press/releases/2009/08/dogs-think.
20. Gregg Vanderheiden and David Kelso, "Comparative Analysis of Fixed-Vocabulary Communication Acceleration Techniques," *Augmentative and Alternative Communication* 3, no. 4 (1987): 196–206. https://doi.org/10.1080/07434618712331274519.
21. John Halloran and Cindy Halloran, *LAMP: Language Acquisition through Motor Planning* (Wooster, OH: The Center for AAC and Autism, 2006).
22. Halloran and Halloran, *LAMP.*
23. Halloran and Halloran, *LAMP.*
24. Jennifer J. Thistle, Stephanie A. Holmes, Madeline M. Horn, and Alyson M. Reum, "Consistent Symbol Location Affects Motor Learning in Preschoolers Without Disabilities: Implications for Designing Augmentative and Alternative Communication Displays," *American Journal of Speech-Language Pathology* 27, no. 3 (2018): 1010–17. https://doi.org/10.1044/2018_ajslp-17-0129.
25. Juliann Woods, Shubha Kashinath, and Howard Goldstein, "Effects of Embedding Caregiver-Implemented Teaching Strategies in Daily Routines on Children's Communication Outcomes," *Journal of Early Intervention* 26, no. 3 (April 1, 2004): 175–93. https://doi.org/10.1177/105381510402600302.
26. Joshua Becker, "Display What You Value Most," Becoming Minimalist, October 2, 2019. https://www.becomingminimalist.com/benefit-display-what-you-value-most/.
27. Saul Mcleod, "Maslow's Hierarchy of Needs," Simply Psychology, March 20, 2020. https://www.simplypsychology.org/maslow.html.
28. Jennifer Kent-Walsh, Kimberly A. Murza, Melissa D. Malani, and Cathy Binger, "Effects of Communication Partner Instruction on the Communication of Individuals Using AAC: A Meta-Analysis." *Augmentative and Alternative Communication* 31, no. 4 (2015): 271–84. https://doi.org/10.3109/07434618.2015.1052153.
29. Rossetti, *The Rossetti Infant Toddler Language Scale.*
30. Halloran and Halloran, *LAMP.*

31. “Augmentative and Alternative Communication Decisions,” American Speech-Language-Hearing Association (ASHA), accessed September 9, 2020. https://www.asha.org/public/speech/disorders/CommunicationDecisions/.

32. Daniel H. Pink, *Drive: The Surprising Truth About What Motivates Us* (Edinburgh, UK: Canongate Books Ltd., 2018).

33. Pink, *Drive.*

34. Pink, *Drive.*

35. David Crystal, “Roger Brown, A First Language: The Early Stages. Cambridge, MA: Harvard University Press, 1973. Pp. Xi 437,” *Journal of Child Language* 1, no. 2 (1974): 289–307. https://doi.org/10.1017/s030500090000074x.

36. Nancy J. Scherer and Lesley B. Olswang, “Role of Mothers’ Expansions in Stimulating Children’s Language Production,” *Journal of Speech, Language, and Hearing Research* 27, no. 3 (1984): 387–96. https://doi.org/10.1044/jshr.2703.387.

37. Rossetti, *The Rossetti Infant Toddler Language Scale.*

38. Halloran and Halloran, *LAMP.*

39. Erika Hoff, *Language Development* (Belmont, CA: Wadsworth/Thomson Learning, 2005).

40. Julia Riedel, Katrin Schumann, Juliane Kaminski, Josep Call, and Michael Tomasello, “The Early Ontogeny of Human–Dog Communication,” *Animal Behaviour* 75, no. 3 (2008): 1003–14. https://doi.org/10.1016/j.anbehav.2007.08.010.

41. Vanderheiden and Kelso, “Comparative Analysis.”

42. Shari Robertson, “Celebrating the Bold Thinking of ASHA Imaginologists,” *The ASHA Leader* 24, no. 11 (2019): 8–10. https://doi.org/10.1044/leader.ftp.24112019.8.